Atmospheric Chemistry

Models and Predictions for Climate and Air Quality

Christine S. Sloane
Thomas W. Tesche
EDITORS

LEWIS PUBLISHERS

Library of Congress Cataloging-in-Publication Data

Atmospheric chemistry: models and predictions for climate and air quality / edited by Christine S. Sloane, Thomas W. Tesche.
p. cm.
Based on presentations at the 198th American Chemical Society Meeting, held Sept. 1989 in Miami Beach, Fla.
Includes bibliographical references and index.
ISBN 0-87371-310-9
1. Atmospheric chemistry—Mathematical models. 2. Climatology—Mathematical models. 3. Air quality—Mathematical models. 4. Air—Pollution—Meteorological aspects—Mathematical models. I. Sloane, Christine S. II. Tesche, T. W. III. American Chemical Society. Meeting (198th : 1989 : Miami Beach, Fla.)

QC879.6.A86 1991
551.5'11—dc20 90-26249
CIP

LEWIS PUBLISHERS, INC.
121 South Main Street, Chelsea, Michigan 48118

PRINTED IN THE UNITED STATES OF AMERICA

Preface

Almost every day we hear of another environmental catastrophe. Some are emergencies; others, like urban smog and acid rain, take a chronic toll; and still others, like the ozone hole and global warming, are harbingers of the future. Each problem must be traced through the labyrinth of atmospheric transport and chemical reactions to a variety of emission sources. Each problem has a unique scale in space and time, in the complexity of the controlling chemical processes, in the strength of feedback mechanisms, and in the economic cost of the relevant pollution control strategies.

Scientists are challenged to find workable solutions to these environmental problems. Ideally, they would design and test a variety of pollution control strategies. However, since each control strategy involves some change in our use of energy or materials, the economic and social cost is too high for full-scale experimentation. Therefore, mathematical models are used to simulate the environmental response to changes in emissions. These models serve as repositories of our collective knowledge of chemical and physical processes in the atmosphere. The modeled simulations of emission control strategies provide the "guinea pigs" that test environmental remedies.

The predictive strength of state-of-the-science mathematical models currently applied to our most pressing atmospheric problems is the subject of this book. It was drawn from a gathering of modeling experts who were at the American Chemical Society meeting in September 1989, where the progress of atmospheric modeling was assessed. The presentations first addressed the large-scale models for ozone depletion and global warming and then moved to the regional-scale models for urban smog (ozone and visibility impairment) and acid rain and the accompanying models of cloud processes and biofeedbacks. Those presentations have been collected for this book.

Christine S. Sloane
T. W. Tesche

Christine S. Sloane is a Senior Staff Research Scientist in the Environmental Science Department at the General Motors Research Laboratory, Warren, Michigan, where she manages the Manufacturing Emissions Control Section. Her participation in professional organizations includes the American Association for Aerosol Research (1991 Program Chair, Board of Directors, Chair of Atmospheric Aerosols Working Group), Air and Waste Management Association (Chair of Visibility Technical Committee), American Chemical Society (symposium organizer), Sigma Xi, and Phi Beta Kappa. Dr. Sloane has received the General Motors Campbell Award for research excellence. Her research interests include air pollution modeling, toxic emissions, aerosol chemistry and physics, and atmospheric optics. She has participated in the analysis of numerous field studies investigating sources of urban air pollution. Dr. Sloane has a Ph.D. degree in Chemical Physics from the Massachusetts Institute of Technology.

T. W. Tesche, Ph.D., Principal Scientist at Alpine Geophysics, Placerville, California, specializes in urban- and regional-scale photochemical, meteorological, and emissions modeling. His principal research interests include the development and evaluation of photochemical air pollution models and the use of advanced modeling methods in regulatory decision making. Dr. Tesche is a Certified Consulting Meteorologist and is a member of the Modeling Advisory Committee (MAC) of the California Statewide Modeling Center. He also serves as Adjunct Professor of Environmental Engineering at the University of California, Davis, where he instructs graduate level classes in advanced atmospheric modeling.

Contributors

Jeffrey S. Gaffney, Environmental Research Division, Building 203, Argonne National Laboratory, 9700 South Cass Avenue, Argonne, Illinois 60439–4843

Jeremy M. Hales, Atmospheric Sciences Department, Battelle Pacific Northwest Laboratories, Battelle Boulevard, Richland, Washington 99352

Prakash Karamchandani, ENSR Consulting and Engineering, 1220 Avenida Acaso, Camarillo, California 93012

Malcolm K. W. Ko, Atmospheric and Environmental Research, Inc., 840 Memorial Drive, Cambridge, Massachusetts 02139

Gürkan Kuntasal, ENSR Consulting and Engineering, 1220 Avenida Acaso, Camarillo, California 93012

Nancy A. Marley, Environmental Research Division, Building 203, Argonne National Laboratory, 9700 South Cass Avenue, Argonne, Illinois 60439–4843

Dennis McNally, Numerical Solutions, 8179 Heather Grove Court, Sacramento, California 95828

Mario J. Molina, Department of Earth, Atmospheric and Planetary Sciences and Department of Chemistry, Massachusetts Institute of Technology, Cambridge, Massachusetts 02139

Stephen H. Schneider, National Center for Atmospheric Research, P.O. Box 3000, Boulder, Colorado 80307–3000

Christian Seigneur, ENSR Consulting and Engineering, 1320 Harbor Bay Parkway, Alameda, California 94501

Christine S. Sloane, Environmental Science Department, General Motors Research Laboratories, 30500 Mound Road, Warren, Michigan 48090–9055

Nien-Dak Sze, Atmospheric and Environmental Research, Inc., 840 Memorial Drive, Cambridge, Massachusetts 02139

T. W. Tesche, Alpine Geophysics, 5360 Spring Circle, Placerville, California 95667

Anne M. Thompson, Atmospheric Chemistry and Dynamics Branch, Code 916, NASA/Goddard Space Flight Center, Greenbelt, Maryland 20771

Akula Venkatram, ENSR Consulting and Engineering, 1220 Avenida Acaso, Camarillo, California 93012

Contents

CHAPTER 1

CHEMISTRY OF STRATOSPHERIC OZONE DEPLETION

Mario J. Molina

INTRODUCTION

In October 1987 the total ozone column over Antarctica dropped to the lowest levels ever observed anywhere in the atmosphere, with more than 95% of the ozone disappearing at some altitudes (Figure 1). In 1989, the Antarctic ozone levels dropped again drastically.

The leading scientific theories which had been postulated to explain this unusual phenomenon included purely dynamical effects (no chemistry), nitrogen chemistry associated with the solar cycle, and chlorine chemistry associated with man-made chlorofluorocarbons (CFCs). The predictions of the CFC theory are based on recent laboratory studies and computer modeling calculations. This theory is the one most clearly supported by field observations. Hence, it is now generally accepted that the ozone hole is a consequence of the release of CFCs to the environment.

The suggestion that CFCs would affect stratospheric ozone was first made in 1974.[1] Due to their chemical inertness, these compounds are stable in the troposphere, and have residence times of the order of a century. They are destroyed predominantly in the upper stratosphere, at low latitudes, by solar ultraviolet (UV) radiation (Figure 2). Their decomposition products reach the polar stratosphere in the form of HCl (hydrogen chloride) and $ClONO_2$ (chlorine nitrate), species which do not destroy ozone directly. Figure 3 shows a schematic representation of the most important species and chemical reactions in the chlorine family.

CATALYTIC CHAIN REACTIONS

Ozone is made predominately in the upper stratosphere at low latitudes. The first step involves dissociation of molecular oxygen by solar radiation in the

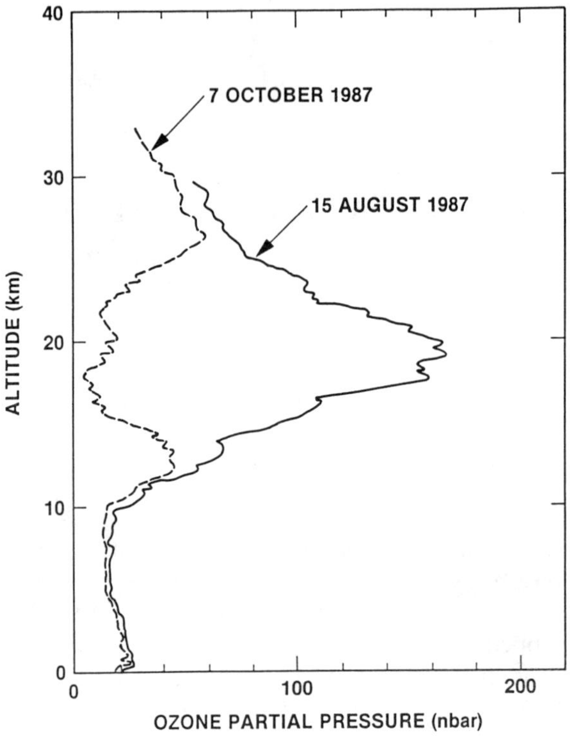

Figure 1. Polar ozone profiles, measured in 1987 over Halley Bay, Antarctica, by Farman.[22]

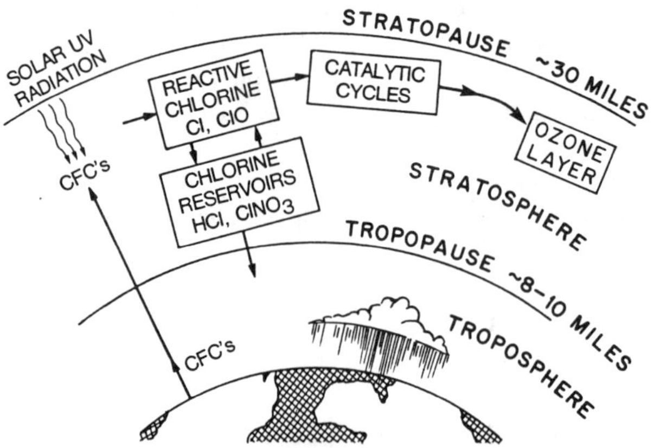

Figure 2. Schematic representation of the main steps in the CFC-ozone depletion theory.

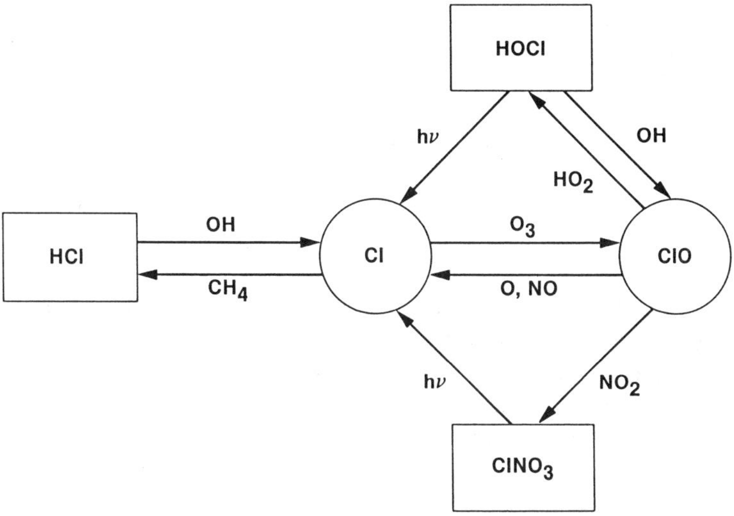

Figure 3. Schematic representation of the key reactions and species in the stratospheric chemistry of chlorine.

Schumann-Runge bands (175 to 205 nm) and the Herzberg continuum (190 to 240 nm) to produce oxygen atoms, which then combine with O_2 to form O_3.

The ozone-destruction chlorine cycles that were first suggested[1,2] require the presence of free oxygen atoms:

$$\begin{array}{r}
Cl + O_3 \rightarrow ClO + O_2 \\
ClO + O \rightarrow Cl + O_2 \\
\hline
\text{Net: } O + O_3 \rightarrow O + O_2
\end{array}$$

Ozone is not formed at high latitudes, and hence free oxygen atoms are scarce in the polar stratosphere. Catalytic ozone-destruction cycles that would operate under these conditions were suggested by Solomon et al.[3] with HOCl as an intermediate, and by McElroy et al.[4] with bromine monoxide as an intermediate:

$$\begin{array}{r}
Cl + O_3 \rightarrow ClO + O_2 \\
ClO + HO_2 \rightarrow HOCl + O_2 \\
HOCl + h\nu \rightarrow OH + Cl \\
OH + O_3 \rightarrow HO_2 + O_2 \\
\hline
\text{Net: } 2O_3 \rightarrow 3O_2
\end{array}$$

$$\begin{array}{r}
ClO + BrO \rightarrow Cl + Br + O_2 \\
Cl + O_3 \rightarrow ClO + O_2 \\
Br + O_3 \rightarrow BrO + O_2 \\
\hline
\text{Net: } 2O_3 \rightarrow 3O_2
\end{array}$$

We suggested that ozone might be destroyed rapidly in the polar stratosphere as a result of the following catalytic cycle:[5]

$$\begin{array}{rcl} ClO + ClO & \rightarrow & Cl_2O_2 \\ Cl_2O_2 + h\nu & \rightarrow & Cl + ClOO \\ ClOO & \rightarrow & Cl + O_2 \\ 2(Cl + O_3 & \rightarrow & ClO + O_2) \\ \hline \text{Net:} \quad 2O_3 & \rightarrow & 3O_2 \end{array}$$

One feature which is common to all of these cycles is that they destroy ozone efficiently only if the NO_2 levels are very low; otherwise, ClO preferentially reacts with NO_2 to form $ClONO_2$. The heterogeneous processes occurring on the polar stratospheric clouds (PSCs) effectively remove NO_2 from the gas phase, by converting it to HNO_3, which remains in the cloud particles.

The species Cl_2O_2, dichlorine peroxide, is the dimer of chlorine monoxide, ClO. There were indications in the literature that such a species would form in the self-reaction of the ClO radical,[6] but its properties were poorly known and even its existence was not well established. In our earlier work[5] we reported infrared (IR) product spectra for systems in which ClO was generated using several different precursors and we concluded that two different isomers of Cl_2O_2 were being formed. We reported the UV spectrum for only one of these isomers, namely, the one formed efficiently with OClO as the excess reactant. Hayman et al.[6] subsequently obtained a different UV spectrum with, however, a similar absorption tail beyond 300 nm. In later work, Cox and Hayman[7] were able to reproduce our UV spectrum and attributed it to a ClO–OClO adduct, rather than to a ClO dimer.

Additional work in our laboratory indicated that this is indeed the case: the two sets of IR bands in our earlier studies correspond to the adduct, Cl_2O_3, and to the dimer, Cl_2O_2. The ClO self-reaction at room temperature generates OClO, and hence this species was always present in our earlier experiments. In our more recent work, by mixing Cl atoms with ozone at much lower temperatures (around 200 K), we were able to produce the dimer in the absence of OClO. Only one of the two sets of bands is present under these circumstances, with a corresponding UV spectrum that is in very good agreement with that reported earlier by Hayman et al.[6] for the dimer, namely, with a maximum around 245 nm.

Additional experiments on the ClO dimer have also been carried out by Burkholder et al.[8] and Friedl and Sander,[9] who reported IR and UV spectra, and by DeMore and Roux,[10] who obtained UV spectra. The results of all these experiments are in very good agreement, and there remains little doubt at present about the spectral characterization, and hence, the existence of the dimer.

The two most likely structures of the dimer are ClOOCl (the symmetric form) and ClOClO (the asymmetric form). Other dimer structures are also possible, e.g., a central Cl atom bonded to each of the other three. Formation of the asymmetric dimer does not lead to a catalytic cycle with net ozone destruction: if it decomposes to produce OClO, subsequent photolysis of this species yields a free oxygen atom which then regenerates an ozone molecule. Recent work by Birk et al.[11] has established by means of submillimeter wave spectroscopy that the structure of the dimer is indeed ClOOCl.

HETEROGENEOUS CHEMISTRY

PSCs form in the cold Antarctic lower stratosphere, providing surfaces on which heterogeneous chemical processes can take place, such as the reaction between HCl and $ClONO_2$ to produce Cl_2 (molecular chlorine), which is released to the gas phase, and HNO_3 (nitric acid), which remains in the solid cloud particles:

$$HCl + ClONO_2 \xrightarrow{\text{ICE}} Cl_2 + HNO_3$$

The Cl_2 molecules absorb light readily, yielding free chlorine atoms, which may initiate catalytic ozone destruction cycles. Solomon et al.[3] were among the first to suggest the reaction between HCl and $ClONO_2$ on PSCs, but such a process was not considered likely because of the low probability for two molecules to collide simultaneously on the surface of a cloud particle. However, our laboratory experiments showed that HCl vapor can be efficiently scavenged by ice-like particles. Furthermore, these experiments indicated that subsequent collisions of $ClONO_2$ molecules with the HCl-containing particles have a high probability of producing Cl_2 and HNO_3.[12] The experimental results of Tolbert et al.[13,14] and of Leu[15,16] confirmed those results and showed that similar heterogeneous reactions involving N_2O_5 (dinitrogen pentoxide) instead of $ClONO_2$ also proceed with high efficiency:

$$HCl + N_2O_5 \xrightarrow{\text{ICE}} NO_2Cl + HNO_3$$

The species NO_2Cl also photolyzes readily, releasing chlorine atoms.

PSCs consist mainly of nitric acid and water, predominately in the form of the acid trihydrate (" type I" clouds). Earlier work showed that the HCl vapor is readily absorbed by pure water ice.[1,17] More recently, our results indicate that this absorption involves the HCl hexahydrate, which is not likely to form under stratospheric conditions because of the small concentrations of HCl present (about 10^{-7} torr). On the other hand, our experiments also show that the nitric acid trihydrate will absorb significant amounts of HCl under such conditions, in agreement with the observations of Hanson and Mauersberger.[18] We now believe that it is the acid-containing, ice-like particles that are particularly active in promoting heterogeneous chemical processes.

This type of heterogeneous chemistry might also occur on sulfuric acid aerosols. However, these aerosols only become active at low temperatures, when they absorb enough water to dissolve significant amounts of HCl vapor.

FIELD OBSERVATIONS

Numerous observations of the chemical composition of the Antarctic stratosphere have shown that the chemistry is highly perturbed, in agreement with expectations based on the processes discussed in the previous sections. The NO_2 levels are found to be extremely low[19] and measurements of ClO by Anderson et al.[20] and by DeZafra et al.[21] have shown that this species is present during the spring months, when ozone is being depleted, at levels of about 1 ppb (Figure 4). These levels are several hundred times larger than predicted in the absence of heterogeneous chemistry and are sufficiently high to explain most, if not all, of the observed ozone disappearance in terms of catalytic cycles involving chlorine.

CONCLUSION

The CFCs are man-made compounds with residence times in the atmosphere of the order of a century. They are destroyed mainly in the stratosphere by

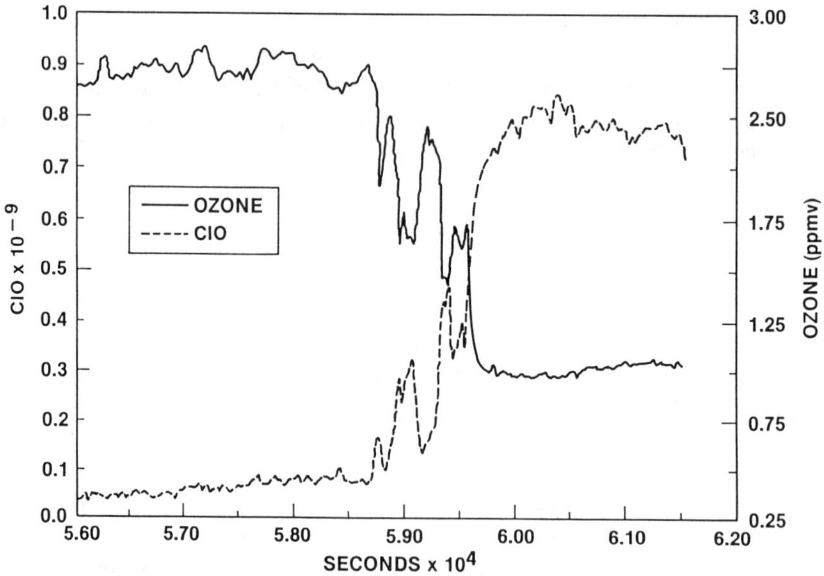

Figure 4. Measurements of chlorine monoxide by Anderson et al.[20] and of ozone by Proffitt et al.[23] carried out in 1987 during the Airborne Antarctic Ozone Experiment (AAOE).

solar UV radiation. Their decomposition products reach the polar statosphere mainly in the form of HCl and $ClONO_2$, and these species react with each other in the presence of PSCs, which consist of ice-like particles. This heterogeneous process has two important consequences: chlorine is released to the gas phase as Cl_2, which photolyzes readily yielding free chlorine atoms, and nitrogen oxides are removed from the gas phase. Catalytic chain reactions involving chlorine free radicals efficiently destroy ozone in the absence of gas phase nitrogen oxides. Field observations have largely corroborated predictions based on this chemistry.

REFERENCES

1. Molina, M.J., and F.S. Rowland. "Stratospheric Sink for Chlorofluoromethanes: Chlorine Atom Catalysed Destruction of Ozone," *Nature* 249:810–814 (1974).
2. Stolarski, R.S., and R.J. Cicerone. "Stratospheric Chlorine: A Possible Sink for Ozone," *Can. J. Chem.* 52:1610–1615 (1974).
3. Solomon, S., R.R. Garcia, F.S. Rowland, and D.J. Wuebbles. "On the Depletion of Antarctic Ozone," *Nature* 321:755–758 (1986).
4. McElroy, M.B., R.J. Salawitch, S.C. Wofsy, and J.A. Logan. "Antarctic Ozone: Reductions Due to Synergistic Interactions of Chlorine and Bromine," *Nature* 321:759 (1986).
5. Molina, L.T., and M.J. Molina. "Production of Cl_2O_2 by the Self Reaction of the ClO Radical," *J. Phys. Chem.* 91:433–436 (1987).
6. Hayman, G.D., J.M. Davies, and R.A. Cox. "Kinetics of the Reaction ClO + ClO → Products and Its Potential Relevance to Antarctic Ozone," *Geophys. Res. Lett.* 13:1347–1350 (1986).
7. Cox, R.A., and G.D. Hayman. "The Stability and Photochemistry of Dimers of the ClO Radical and Implications for Antarctic Ozone Depletion," *Nature* 322:796–800 (1988).
8. Burkholder, J.B., J.J. Orlando, and C.J. Howard. "Ultraviolet Absorption Cross Sections of Cl_2O_2 between 210 and 410 nm," *J. Phys. Chem.* 94:687–695 (1990).
9. Friedl, R.R., and S.P. Sander. "Studies of ClO and BrO Reactions Important in the Polar Stratosphere: Kinetics and Mechanism in the ClO + BrO and ClO + ClO Reactions," paper presented at the Polar Ozone Workshop, Snowmass, CO, May 9 to 13, 1988.
10. DeMore, W.B., and E.T. Roux. "Chemistry of the ClO Dimer at Low Temperatures," paper presented at the Polar Ozone Workshop, Snowmass, CO, May 9 to 13, 1988.
11. Birk, M., R.R. Friedl, E.A. Cohen, H.M. Pickett, and S.P. Sander. "The Rotational Spectrum and Structure of Chlorine Peroxide," *J. Chem. Phys.* 91:6588–6597 (1989).
12. Molina, M.J., T.L. Tso, L.T. Molina, and F. C.-Y. Wang. "Antarctic Stratospheric Chemistry of Chlorine Nitrate, Hydrogen Chloride and Ice: Release of Active Chlorine," *Science* 238:1253–1257 (1987).
13. Tolbert, M.A., M.J. Rossi, R. Malhotra, and D.M. Golden. "Reaction of Chlorine Nitrate with Hydrogen Chloride and Water at Antarctic Stratospheric Temperatures," *Science* 238:1258–1260 (1987).

14. Tolbert, M.A., M.J. Rossi, and D.M. Golden. "Antarctic Ozone Depletion Chemistry: Reactions of N_2O_5 with H_2O and HCl on Ice Surfaces," *Science* 240:1018 (1988).
15. Leu, M.T. "Laboratory Studies of Sticking Coefficients and Heterogeneous Reactions Important in the Antarctic Stratosphere," *Geophys. Res. Lett.* 15:17–20 (1988).
16. Leu, M.T. "Heterogeneous Reactions of N_2O_5 with H_2O and HCl on Ice Surfaces: Implications for Antarctic Ozone Depletion," *Geophys. Res. Lett.* 15:855–858 (1988).
17. Wofsy, S.C., M.J. Molina, R.J. Salawitch, L.E. Fox, and M.B. McElroy. "Interactions between HCl, NO_x, and H_2O Ice in the Antarctic Stratosphere: Implications for Ozone," *J. Geophys. Res.* 93:2442–2450 (1988).
18. Hanson, D.R., and K. Mauersberger. "Solubility and Equilibrium Vapor Pressures of HCl Dissolved in Polar Stratospheric Cloud Materials: Ice and Trihydrate of Nitric Acid." *Geophys. Res. Lett.* 15:1507–1510 (1989).
19. Fahey, D.W., D.M. Murphy, K.K. Kelly, M.K.W. Ko, M.H. Proffitt, C.S. Eubank, G.V. Ferry, M. Loewenstein, and K.R. Chan. "Measurements of Nitric Oxide and Total Reactive Nitrogen in the Antarctic Stratosphere: Observations and Chemical Implications," *J. Geophys. Res.* 94:16,665–16,681 (1989).
20. Anderson, J.G., W.H. Brune, S.A. Lloyd, D.W. Toohey, S.P. Sander, W.L. Starr, M. Loewenstein, and J.R. Podolske. "Kinetics of Ozone Destruction by ClO and BrO within the Antarctic Vortex: An Analysis Based on In Situ ER-2 Data," *J. Geophys. Res.* 94:11,480–11,520 (1989).
21. DeZafra, R.L., M. Jaramillo, A. Parrish, P. Solomon, B. Connor, and J. Barnett. "Observation of Abnormally High Concentrations of Chlorine Monoxide at Low Altitudes in the Antarctic Spring Stratosphere. I. Diurnal Variation," *Nature* 328:408 (1987).
22. Farman, J.C. "Ozone Observations at Halley Bay, 1987, " paper presented at the Polar Ozone Workshop, NASA, Snowmass, CO, May 9 to 13, 1988.
23. M.H. Proffitt, M.J. Steinkamp, J.A. Powell, R.J. McLaughlin, O.A. Mills, A.L. Schmeltekopf, T.L. Thompson, A.F. Tuck, T. Tyler, R.H. Winker, and K.R. Chan. "In Situ Ozone Measurements within the 1987 Antarctic Ozone Hole from a High-Altitude ER-2 Aircraft," *J. Geophys. Res.* 94:16,547–16,555 (1989).

CHAPTER 2

Impact on Stratospheric Ozone from Emissions of CFCs and Substitutes: a Two-Dimensional Model Assessment

Malcolm K. W. Ko and Nien-Dak Sze

INTRODUCTION

The spatial and temporal distributions of ozone (O_3) in the stratosphere reflect complex interactions among radiation, chemistry, and dynamics. Current findings indicate that human activities may perturb the natural state of the environment resulting in a decrease of O_3 in the future atmosphere. A decrease in O_3 would increase the amount of harmful ultraviolet (UV) radiation reaching the earth surface causing adverse impact on the biosphere.[1]

Predictions from computer model simulations provide the only means to assess the response of O_3 to future changes in the environment. The reliability of these model predictions are judged by how well the models simulate the present-day atmosphere. However, validation of model results is not straightforward. Models can at best simulate the climatological mean state of the atmosphere. Because of natural variabilities in the atmosphere, it is difficult to judge how well a model is performing by comparing the model output with a particular snapshot of the atmosphere. In some cases, it is not clear how long a database is needed to define a climatological mean state.

Over the years, different kinds of models have had a varying degree of success in simulating the behavior of O_3. Among the hierarchy of models, the two-dimensional (2-D) (latitude and height) models provide the best compromise between flexibility in performing multi-year simulations and yet producing results with latitudinal and seasonal variations that could be meaningfully compared with the vast amount of satellite data. A number of independent efforts funded by government and industry have made remarkable progress in the past decades.[2-14] Results from many of these models were reviewed recently in the workshop on Two-dimensional Intercomparison of Stratospheric Models.[15] These models have been accepted as viable assessment tools in the most recent UNEP/WMO Report.[16]

MODEL SIMULATION OF ATMOSPHERIC TRACE GASES

The processes that control the concentrations of trace gases in the atmosphere as well as their interactions are presented schematically in Figure 1. The exchange of gases between the atmosphere and the pedosphere is represented in the model by the boundary conditions that specify either the surface concentrations of the gases or the amount of material introduced into or removed from the atmosphere. The spatial and seasonal distributions of the trace gases in the atmosphere are affected by chemical transformations as well as large-scale transport that carries the gases from one region to another. Depending on the design of the model, the transport circulation and temperature may be prescribed or calculated within the model itself. In this chapter, we will restrict our discussion to the results from models in which the annual cycle of transport circulation and temperature are fixed.

The results from the AER (Atmospheric and Environmental Research, Inc.) model are presented here. The AER two dimensional (2-D) model was developed over the past 10 years with funding from both government and private agencies. The model has been used to assess the impacts on O_3 due to emissions of halocarbons;[1,16,17] to high-altitude aircraft operations;[18] to atmospheric nuclear testing;[19] to global increase of trace gases such as CO_2, CH_4 and N_2O;[16] to the dilution effect due to the antarctic O_3 hole;[20] to heterogeneous chemistry occurring on the global aerosol layer;[21] and most recently to the emissions of CFC substitutes.[22,23]

Description of the AER model is given in previous publications.[7,10,17] A brief description of the key elements is given here. The model simulates the seasonal, latitudinal, and vertical distribution of atmospheric trace gases by

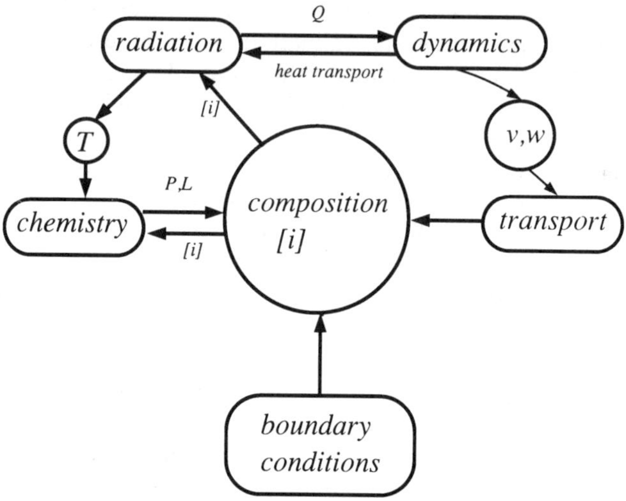

Figure 1. Schematic diagram illustrating the processes that control the concentrations of trace gases in the atmosphere.

Table 1. Summary of Chemical Species Included in AER 2-D Model

Fixed species	
M	Bulk air density calculated from ideal gas law
O_2	Set equal to 21% of M
H_2O	Stratospheric values parameterized from the LIMS observation[33]
	Tropospheric values calculated from seasonally varying relative humidity determined from climatology
Long-lived species	
N_2O	
CH_4, H_2, CO	
CH_3Cl, CH_3CCl_3, CCl_4	
CFC-11, CFC-12, CFC-22	
CF_2O, CFClO, CCl_2O	
CH_3Br, Halon-1301, Halon-1211	
NOY = N + NO + NO_2 + NO_3 + 2 × N_2O_5 + HNO_3 + $ClNO_3$ + HNO_4	
ClY = Cl + ClO + $ClNO_3$ + HCl + HOCl + OClO + 2 × Cl_2O_2	
BrY = Br + BrO + $BrNO_3$ + HBr + HOBr + 2 × Br_2	
O_x = O + O(^{1}D) + O_3	
Radical species	
Oxygen	O, O(^{1}D)
Hydrogen	H, OH, HO_2, H_2O_2
Methyl	CH_3O_2, CH_2O, CH_3OOH
Nitrogen	N, NO, NO_2, NO_3, N_2O_5, HNO_3, $ClNO_3$, HNO_4
Chlorine	Cl, ClO, $ClNO_3$, HCl, HOCl, OClO, Cl_2O_2
Bromine	Br, BrO, $BrNO_3$, HBr, HOBr, Br_2

accounting for chemical interactions among the gases and dynamical transport. The model covers the globe from pole to pole with a latitudinal resolution of 9.5 degrees. Vertical coverage is from the ground to approximately 60 km in a log-pressure grid. The vertical resolution is approximately 3.5 km.

The model includes approximately 40 chemical species which interact through over 100 chemical reactions. The kinetic reaction rates are those given by the NASA review panel.[24] The solar flux and absorption cross sections are from the WMO/NASA report.[1] The chemical scheme employs the grouping technique to deal with chemicals having vastly different atmospheric lifetimes. Table 1 shows the species in the model classified according to their lifetimes. The radical species have short lifetimes and are calculated assuming photochemical equilibrium. Their concentrations are allowed to vary diurnally. Long-lived species, which include the source gases, are calculated using the mass-continuity equation.

The temperature in the atmosphere is specified as a function of latitude, height, and season with values determined from climatology. Dynamical transport within the model is effected by the zonal-mean circulation and turbulent mixing parameterized by quasi-horizontal diffusion along isentropic surfaces, and by vertical diffusion in the troposphere and upper stratosphere. The model circulation was derived from diabatic heating rates based on those calculated by Murgatroyd and Singleton[25] for the upper stratosphere and by Dopplick[26] for the lower stratosphere.

The horizontal eddy diffusion coefficient, K_{yy}, for the troposphere and lower stratosphere varies from 3×10^9 cm²/sec at low latitudes to 6×10^9 cm²/sec at mid-latitudes in fall and winter or 2×10^{10} cm²/sec at mid-latitudes in spring and summer. These values yield a good fit to observed ozone profiles in the lower stratosphere and are close to the magnitudes derived by Newman and Schoeberl.[27] The value of K_{yy} in the stratosphere above 25 km is 3×10^9 cm²/sec for all latitudes and seasons. This is based on the work of Kida[28] and Tung[29] in estimating an average horizontal diffusion coefficient for the stratosphere.

The vertical diffusion coefficient, K_{zz}, is 1×10^5 cm²/sec in the troposphere, 1×10^3 cm²/sec in the stratosphere below 40 km, and 1×10^4 cm²/sec above 40 km. Stratospheric vertical diffusion was estimated by Kida[28] to be 1×10^3 cm²/sec. Enhanced vertical mixing above 40 km is based on the work of Garcia and Solomon[30] regarding gravity wave breaking.

Concentrations of long-lived atmospheric species are integrated forward in time, with the change in mixing ratio per unit time in the latitude (ϕ)-log-pressure (s) coordinate system given by:

$$\frac{\partial f}{\partial t} = -\frac{1}{a\cos\phi}\frac{\partial}{\partial\phi}(fv\cos\phi) - e^{s}\frac{\partial}{\partial s}(fwe^{-s}) + \frac{1}{a^2\cos\phi}\frac{\partial}{\partial\phi}\left[K_{yy}\cos\phi\frac{\partial f}{\partial\phi}\right] + e^{s}\frac{\partial}{\partial s}\left[K_{zz}e^{-s}\frac{\partial f}{\partial s}\right] + P - Lf$$

where f represents the zonal-mean volume mixing ratio of a trace species; a is the radius of the earth, v is the zonal-mean horizontal transport velocity, w is the zonal-mean vertical transport velocity, and K_{yy} and K_{zz} are the horizontal and vertical eddy mixing coefficients. Chemical production and loss are represented by the terms P and Lf, respectively. The finite differencing scheme used is that developed by Smolarkiewicz.[31] It is an iterative upstream scheme which removes much of the implicit diffusion of upwind differencing by adding a corrective step to each time step. The scheme computes fluxes at grid box boundaries and transports mass only in the direction of fluid flow. Negative mixing ratios are not generated provided that the time step is small enough.

The role of the trace gases in determining ozone is highlighted in Figure 2. As the source gases from the ground are transported to the stratosphere, they are dissociated to form the respective radical species. Photochemical reactions between O_3 and the radical species help to determine the concentration of O_3. As the concentration of the source gases changes, it tends to change the concentration of O_3 through their impacts on the radical species.

The total column ozone generated by the AER model for the present-day atmosphere is shown in Figure 3(A) as a function of latitude and time of the year. Figure 3(B) shows the corresponding observed ozone column abundance. The model successfully simulates both the magnitude of the observed ozone column and its seasonal and latitudinal variation. The ozone column shows a minimum near the equator and a maximum at high latitudes in the springtime.

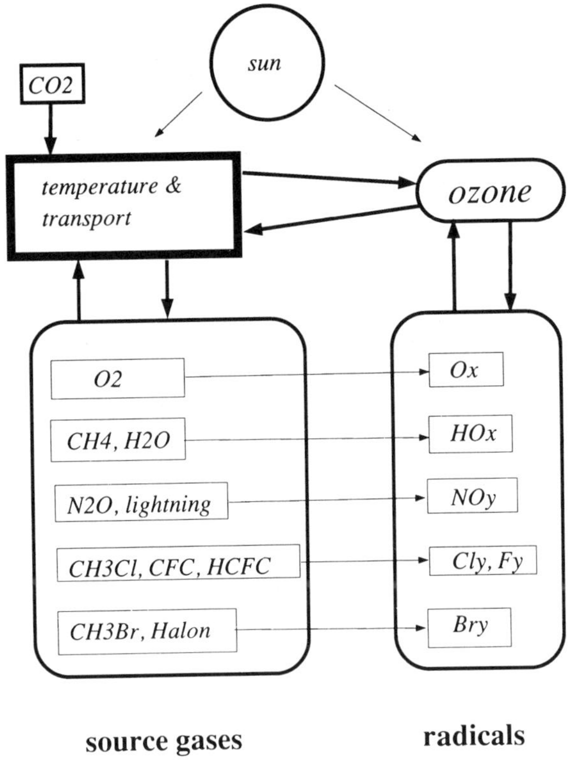

Figure 2. The roles of the source gases and the radical species in determining the concentration of ozone in the atmosphere.

Since ozone is generated primarily near the equator by photolysis of molecular oxygen, this indicates the importance of transport in determining the global ozone distribution. The calculated ozone mixing ratios as functions of latitude and height for January, April, July, and October are shown in Figure 4. Ozone profiles derived from satellite observations are shown in Figure 5 for comparison.

EFFECT OF CHLOROFLUOROCARBONS (CFCs) AND HYDROGENATED HALOCARBONS (HCFCs) ON O_3

The calculation of the response of O_3 to CFCs emissions depends on the simulations of a series of physical processes illustrated in Figure 6. As first noted by Molina and Rowland,[32] chlorine atoms released by photochemical dissociation of the CFCs (and HCFCs) could participate in the catalytic removal of ozone. Two examples of such cycles are

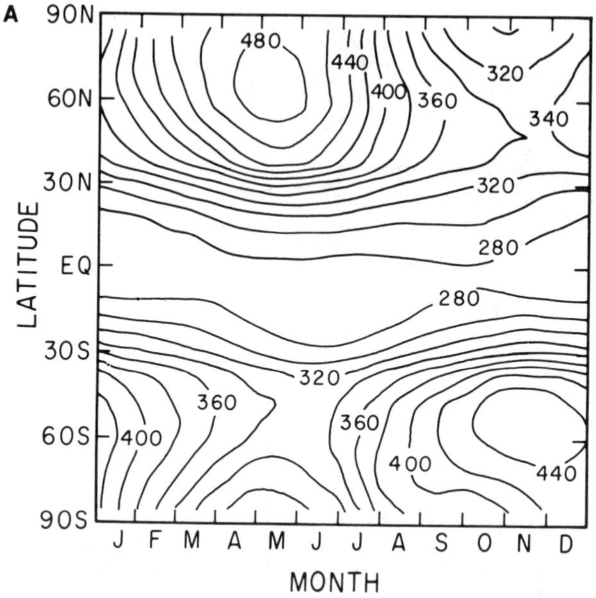

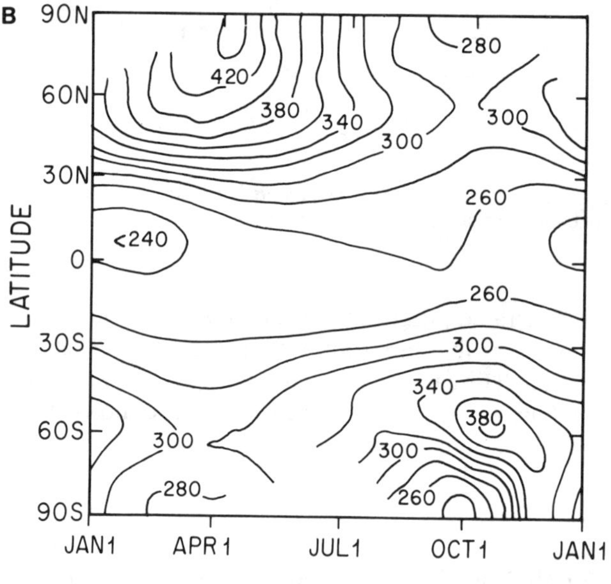

Figure 3. Total ozone column (Dobson units) as a function of latitude and time of the year (A) as calculated by the AER 2-D chemical-transport model for present-day atmospheric conditions and (B) derived from Nimbus 7 satellite observations[34] for the years 1978 to 1982. Observations have been averaged into zonal means and smoothed by computing 10-day averages. (Figure 3(B) reprinted from Bowman, K. P. and A. J. Krueger. "A Global Climatology of Total Ozone from the Nimbus 7 Total Ozone Mapping Spectrometer," *J. Geophys. Res.* 90:7967–7976 (1985).)

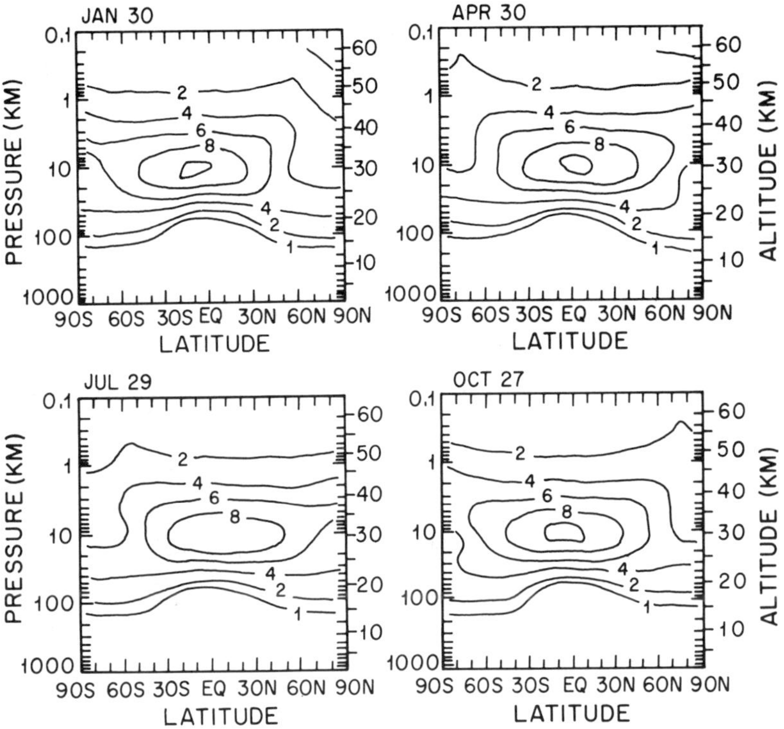

Figure 4. Model calculated cross sections of ozone mixing ratio (ppmv) as a function of latitude and height for January, April, July, and October for present-day conditions. Contours are 1, 2, 4, 6, 8, and 10 ppmv.

$$\begin{array}{l} Cl + O_3 \rightarrow ClO + O_2 \\ ClO + O \rightarrow Cl + O_2 \\ \hline O + O_3 \rightarrow 2O_2 \end{array}$$

$$\begin{array}{rcl} 2(Cl + O_3 & \rightarrow & ClO + O_2) \\ ClO + ClO + M & \rightarrow & Cl_2O_2 + M \\ Cl_2O_2 + h\nu & \rightarrow & Cl + ClO_2 \\ ClO_2 + M & \rightarrow & Cl + O_2 + M \\ \hline O_3 + O_3 & \rightarrow & 3O_2 \end{array}$$

Note that the role of the chlorine atoms is that of a catalyst. Thus, an atom, once released, can participate in the O_3 removal as long as it stays in the atmosphere. Chlorine atoms released in the stratosphere will have a larger impact since most of the O_3 resides in the stratosphere and the residence time for chlorine atoms is about 2 years in the stratosphere before they are transported to the troposphere. In contrast, chlorine atoms in the troposphere are rapidly removed and have little impact on ozone. The effect of CFCs on O_3 is large, precisely because they are inert in the troposphere so that they are dissociated in the stratosphere and deposit the chlorine atoms in the lower stratosphere where they have the largest impact. In contrast, the HCFCs are

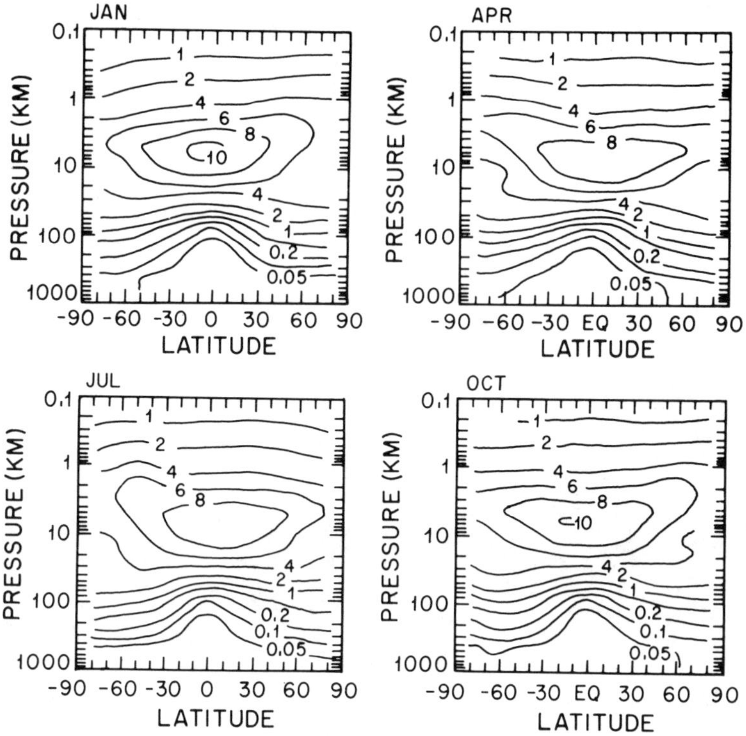

Figure 5. Observed cross sections of ozone mixing ratio (ppmv) from the Nimbus 7 satellite as a function of latitude and height. Shown are monthly averages for January, April, July, and October 1979, averaged by 10–degree latitude bands.[35] Contours are 0.05, 0.1, 0.2, 0.5, 1, 2, 4, 6, 8, and 10 ppmv. (Reprinted from McPeters, R. D., D. F. Heath, and P. K. Bhartia. "Average Ozone Profiles for 1979 from the Nimbus 7 SBUV Instrument," *J. Geophys. Res.* 89: 5199–5214 (1984).)

chosen as substitutes exactly because much of them is removed in the troposphere.

Results from typical scenario calculations are presented below. These scenarios were among those discussed in the UNEP/WMO Report.[16] Current findings suggest that both CH_4 and N_2O are increasing with time at the rate of 16 ppbv/year and 0.25%/year, respectively. In the calculation, it is assumed that they will continue to increase at these rates. At the same time, emissions of CFC and Halons are expected to continue. In addition, the concentration of CCl_4 is assumed to increase at the rate of 1 pptv/year from its 1985 concentration of 100 pptv, while that of CH_3CC1_3 is assumed to increase at 4 pptv/year from its 1985 value of 130 pptv. Finally, the emission rate of CFC-22 is assumed to show a 7 kiloton/year increase over its 1985 value of 140 kiloton/year. The combined effects of these trace gas changes can be incorporated into the boundary conditions for a model run. The resulting percent change in O_3 column abundance and local concentration by the year 2060 is shown in

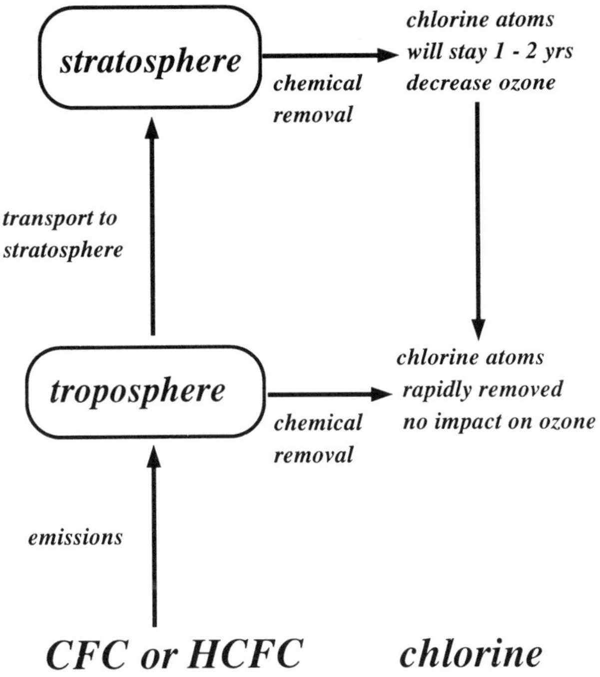

Figure 6. Schematic diagram showing the fate of the CFCs and HCFCs and their impact on ozone.

Figure 7. An alternate calculation was performed to see the effect of using a substitute in place of the CFCs. In the calculation, a 50% cut back in CFC-11, CFC-12, and Halons is assumed by the year 2000. The compound HCFC-22 is used in place of CFC-11 and CFC-12, where 1 lb of HCFC–22 replaces 2 lb of CFC-11 or CFC-12. The calculated O_3 change is shown in Figure 8. Comparison of Figures 7 and 8 shows that replacing CFC-11 and CFC-12 with CFC-22 would lead to a reduction of the calculated O_3 decrease by three percentage points at high latitudes. However, the projected chlorine content in the year 2060 is still considerably higher than the 1980 value and the O_3 decrease persists. It should be noted that the projected emissions in both cases considered here are larger than those discussed in the London Amendment of the Montreal Protocol.

CONCEPT OF OZONE DEPLETION POTENTIAL (ODP)

An estimate of the impact of a substitute can be obtained by calculating the ozone depletion potential (ODP) which is a measure of the steady-state impact of a species on global O_3 content relative to CFC-11 for the same emission rates.[23] Most of the substitutes under consideration are hydrogenated

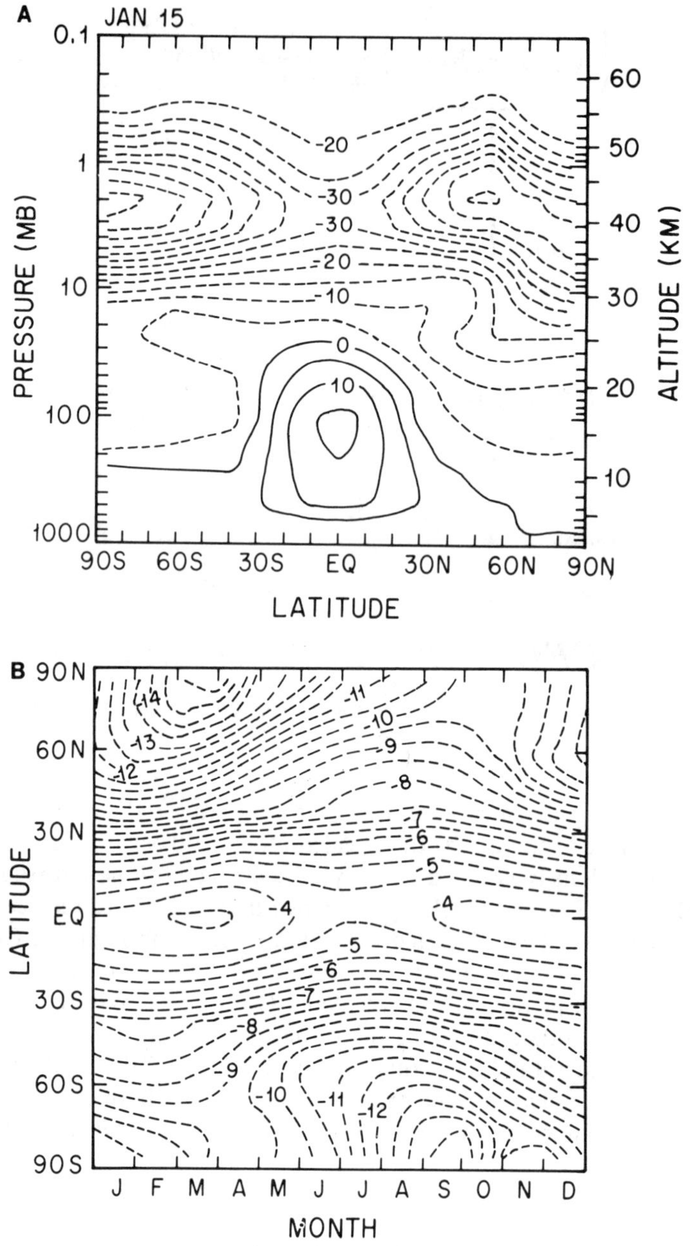

Figure 7. Calculated percent changes in O_3 between the years 1980 and 2060 for scenario A. Panel A shows the percent change as a function of latitude and altitude for January. The percent change in column abundance as a function of latitude and season is shown in panel B.

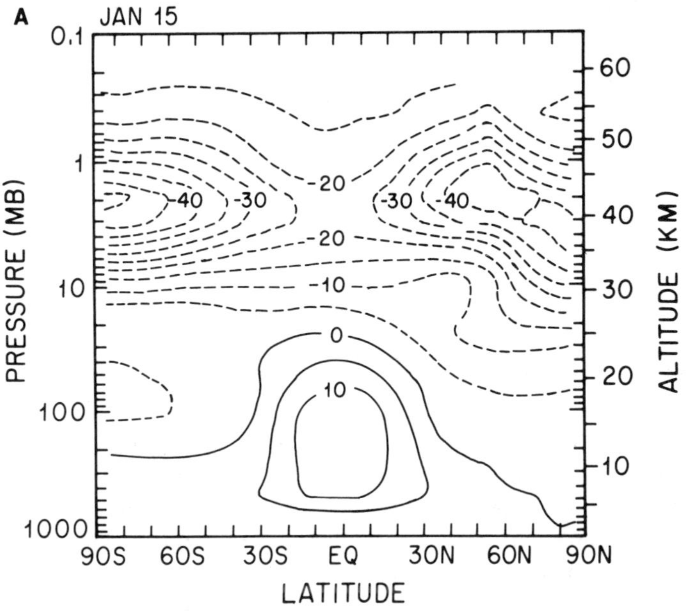

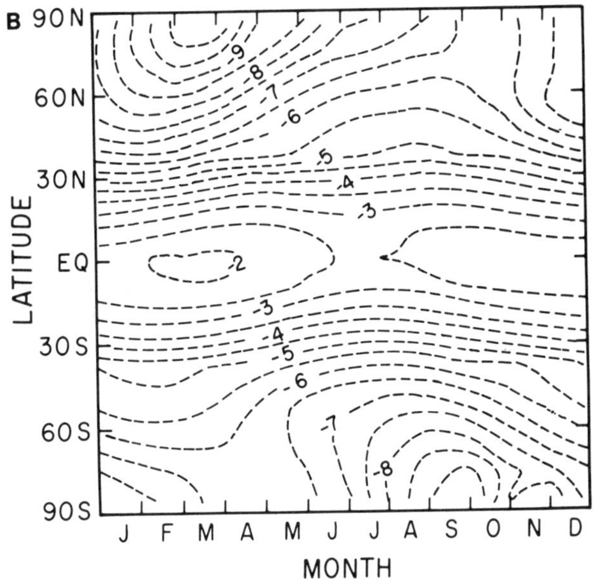

Figure 8. Calculated percent changes in O_3 between the years 1980 and 2000 for scenario B. Panel A shows the percent change as a function of latitude and altitude for January. The percent change in column abundance as a function of latitude and season is shown in panel B.

Table 2. Model Calculated Ozone Depletion Potential (ODP) for Selected Species

	Lifetime (years)	ODP
F-11	47	1.0
F-12	95	0.88
F-115	399	0.37
F-22	24	0.07
F-123	2.4	0.027
F-124	10	0.030
F-125	43	—*
F-134a	24	—*
F141b	11	0.14
F142b	28	0.077
F152a	2.7	—*

Note: Asterisk (*) indicates molecules do not contain any chlorine atoms.

halocarbons (HCFCs) which react with the radical OH in the troposphere. As a result, a much smaller portion of the chlorine will enter the stratosphere. Values for ODP of some of the species that have been identified as potential substitutes are given in Table 2. Note that the values are typically 0.1 or smaller, i.e., the steady-state impact on global O_3 content is one tenth that of CFC-11.

While ODP value is useful in providing a single number to characterize the O_3 impact from a species, its limitations must be recognized. The expected O_3 decrease resulting from the atmospheric release of each species is a function of latitude and season. A single number may not be adequate to characterize the impact on the biosphere. The ODP value, which represents the steady state impact, does not provide information on the transient response in the 40- to 50-year time frame important for regulation purposes. Finally, the effects of special dynamics in the polar region and the role of heterogeneous chemistry occurring on the polar stratospheric cloud and/or the global sulfate aerosol layer have not been included in current calculations and must be evaluated.[36,37] The dependence of ODP on OH concentrations in the troposphere results in additional coupling between the response of stratospheric O_3 and future global changes. Anticipated surface warming due to CO_2 increase and changes in other trace gases is expected to change the atmospheric concentration of OH. Such changes will also modify the response of O_3 to emissions of the HCFCs.

ACKNOWLEDGMENT

The modeling work at AER has been funded by the Fluorocarbon Program Panel of the Chemical Manufacturers Association, the Upper Atmosphere Theory and Data Analysis Program of the National Aeronautics and Space Administration, and the U.S. Air Force Geophysics Laboratory. Work on CFC substitutes is also funded by the Alternative Fluorocarbon Environmental Acceptability Study (AFEAS).

REFERENCES

1. "Atmospheric Ozone: Assessment of Our Understanding of the Processes Controlling Its Present Distribution and Changes," WMO/NASA, World Meteorological Organization, Global Ozone Research and Monitoring Project, Report No. 16, Geneva (1986).
2. Harwood, R. S. and J. A. Pyle. "A Two-Dimensional Mean Circulation Model for the Atmosphere Below 80 km," *Quart. J. Roy. Soc.* 101:723–747 (1975).
3. Vupputuri, R. K. R. "The Structure of the Natural Stratosphere and the Impact of Chlorofluoromethanes on the Ozone Layer Investigated in a 2-D Time Dependent Model," *Pure Appl. Geophys.* 117:448–485 (1979).
4. Pyle, J. A. and C. F. Rogers. "A Modified Diabatic Circulation Model for Stratospheric Tracer Transport," *Nature* 287:711–714 (1980).
5. Miller, C., D. L. Filkin, A. J. Owens, J. M. Steed, and J. P. Jesson. "A Two–Dimensional Model of Stratospheric Chemistry and Transport," *J. Geophys. Res.* 86:12,039–12,065 (1989).
6. Garcia, R. R. and S. Solomon. "A Numerical Model of the Zonally Average Dynamical and Chemical Structure of the Middle Atmosphere," *J. Geophys. Res.* 88:1379–1400 (1983).
7. Ko, M. K. W., N. D. Sze, M. Livshitz, M. B. McElroy, and J. A. Pyle. "Seasonal and Latitudinal Behaviors of Trace Gases and O_3 as Simulated by a 2-D Model of the Atmosphere," *J. Atmos. Sci.* 41:2381–2408 (1984).
8. Guthrie, P. D., C. H. Jackman, J. R. Herman and C. J. McQuillan. "A Diabatic Circulation Experiment in a Two-Dimensional Photochemical Model," *J. Geophys. Res.* 89:9589–9602 (1984).
9. Stordal, F., I. S. Isaksen and K. Horntveth. "A Diabatic Circulation Two–Dimensional Model with Photochemistry: Simulations of Ozone and Ground Released Tracers," *J. Geophys. Res.* 90:5757–5776 (1985).
10. Ko, M. K. W., K. K. Tung, D. K. Weisenstein, and N. D. Sze. "A Zonal-Mean Model of Stratospheric Tracer Transport in Isentropic Coordinates: Numerical Simulations for Nitrous Oxide and Nitric Acid," *J. Geophys. Res.* 90:2313–2329 (1985).
11. Douglass, A. R., C. H. Jackman and R. S. Stolarski. "Comparison of Model Results Transporting the Odd Nitrogen Family with Results Transporting Separate Odd Nitrogen Species," *J. Geophys. Res.* 94:9862–9872 (1989).
12. Brasseur, G., and M. H. Hitchman. "Stratospheric Response to Trace Gas Perturbations: Changes in Ozone and Temperature Distributions," *Science* 241:634–637 (1988).
13. Schneider, H. R., M. K. W. Ko, N. D. Sze, G.-Y. Shi, and W.-C. Wang. "An Evaluation of the Role of Eddy Diffusion in Stratospheric Interactive Two–Dimensional Models," *J. Atmos. Sci.* 46:2079–2093 (1989).
14. Wuebbles, D. J. and D. E. Kinnison. "A Two-Dimensional Study of Past Trends in Global Ozone," Proc. Quadrennial Ozone Symposium, University of Gottingen, Germany (1988).
15. Jackman, C. H., R. K. Seals, Jr., and M. J. Prather, Eds. "Two-Dimensional Intercomparison of Stratospheric Models," NASA Conference Publication 242, Virginia Beach, VA (September 1988).
16. "Scientific Assessment of Stratospheric Ozone: 1989," UNEP/WMO, World Mete-

orological Organization, Global Ozone Research and Monitoring Project, Report No. 20, Geneva (1990).
17. Ko, M. K. W., N. D. Sze, and D. K. Weisenstein. "The Roles of Dynamical and Chemical Processes in Determining the Stratospheric Concentration of Ozone in One-Dimensional and Two-Dimensional Models," *J. Geophys. Res.* 94:9889–9896 (1989).
18. Ko, M. K. W., D. K. Weisenstein, N. D. Sze, J. M. Rodriguez, and C. Heisey. "Effects of Engine Emissions from High Speed Civil Transport Aircraft: A Two-Dimensional Modeling Study," Final Report to ST Systems Corporation, Hampton, VA, Contract No. 88-6209-D1417 (1989).
19. Watson, R. T., M. J. Prather, and M. J. Kurylo. "Present State of Knowledge of the Upper Atmosphere 1988," An Assessment Report, NASA Publication 1208, Washington, D.C. (1988).
20. Sze, N. D., M. K. W. Ko, D. K. Weisenstein, J. J. Rodriguez, R. S. Stolarski, and M. R. Schoeberl. "Antarctic Ozone Hole: Possible Implications for Ozone Trends in the Southern Hemisphere," *J. Geophys. Res.* 94:11,521 (1989).
21. Rodriguez, J. M., Ko, M. K. W., N. D. Sze, D. Weisenstein and H. R. Schneider. "Global Implication of Heterogeneous Chemistry: Assessment Utilizing a Two-Dimensional Model," paper presented at the IAMAP Symposium on Middle Atmosphere Sciences, August 2 to 7, 1989, Reading, U.K. (1989).
22. Ko, M. K. W., N. D. Sze, and D. K. Weisenstein. "Model Evaluation of CFM Replacements," paper presented at The Conference on Atmospheric Chemistry, Boulder, CO, June 18 to 23, 1989.
23. Fisher, D. A., C. H. Hales, D. L. Filkin, M. K. W. Ko, N. D. Sze, P. S. Connell, D. J. Wuebbles, I. S. A. Isaksen, and F. Stordal. "Relative Effects on Stratospheric Ozone of Halogenated Methanes and Ethanes of Social and Industrial Interest," *Nature*, 344:508–512 (1990).
24. "Chemical Kinetics and Photochemical Data for Use in Stratospheric Modeling, Evaluation Number 8," NASA/JPL, JPL Publication 87–41, Pasadena, CA (1988).
25. Murgatroyd, R. J., and F. Singleton. "Possible Meridional Circulation in the Stratosphere and Mesosphere," *Quart. J. Roy. Meteorol. Soc.* 87:125–135 (1961).
26. Dopplick, T. G. "Radiative Heating of the Global Atmosphere: Corrigendum," *J. Atmos. Sci.* 36:1812–1817 (1979).
27. Newman, P. A. and M. R. Schoeberl. "Horizontal Mixing Coefficients for Two-Dimensional Chemical Models Calculated From National Meteorological Center Data," *J. Geophys. Res.* 91:7919–7924 (1986).
28. Kida, H. "General Circulation of Air Parcels and Transport Characteristics Derived from a Hemispheric GCM, 2, Very Long-Term Motions of Air Parcels in the Troposphere and Stratosphere," *J. Meteorol. Soc. Jpn.* 61:510–523 (1983).
29. Tung, K. K. "Modeling of Tracer Transport in the Middle Atmosphere," in *Dynamics of the Middle Atmosphere*, J. R. Holton, and T. Matsuno, Eds., (Tokyo: Terra Scientific Publishing, 1984), pp. 412–444.
30. Garcia, R. R., and S. Solomon. "The Effect of Breaking Gravity Waves on the Dynamics and Chemical Composition of the Mesosphere and Lower Thermosphere," *J. Geophys. Res.* 90:3850 (1985).
31. Smolarkiewicz, P. K. "A Fully Multidimensional Positive Definite Advection Transport Algorithm with Small Implicit Diffusion," *J. Comput. Phys.* 54:325 (1984).

enhouse gases"[2] will enhance the heat-trapping capacity of the atmos-
. re-radiate increased amounts of IR energy toward the earth's sur-
re then is the controversy that the ongoing "geophysical experiment"
? Despite the contentious nature of discussion in the popular press
her the greenhouse effect is or isn't a valid theory, there is, I've said,
o debate in the scientific community over the greenhouse effect as a
proposition. However, there is controversy. The debate is over
ncreases in gases such as CO_2, methane, and chlorofluorocarbons
se the earth's temperature by 1, 5, or 8°C, whether that will happen
, or 150 years, whether it will be drier in Iowa and wetter in India,
iese changes are good or bad and for whom, and ultimately whether
l do anything about it now or in the future. There is plenty of
sy to go around, particularly in the latter area. Therefore it is essen-
ve have a reasonable understanding of what is well known, known
nd not known at all in this subject so punctuated by considerable
y, but also by considerable knowledge.

MATIC FUTURE

l Perspective

: has varied by a substantial amount over the 4.6 billion years of
ory. We know very little about the climate of ancient times before
life left sufficient fossil evidence, but we do know that the very
of life, as well as sedimentary rocks some 3.5 to 4 billion years ago,
at planetary temperatures could not have been too much above 40°C
)°C, at least in some parts of the earth. Some have even hypothesized
was as much as 1000 times more CO_2 in the earth's atmosphere then,
vided adequate warmth to compensate for the 30% or so reduction
iergy radiated by the faint young sun of 4 million years ago. While
theses are intriguing, the paucity of data makes them not very useful
nd of quantitative validation of present estimates of how much the
ht warm given specific increases in greenhouse gases. We could turn
sozoic era, a much warmer period of earth's history, where fossil
s more abundant to suggest that the planet was something like 10 to
mer. This was a time in which great fossil fuel deposits were laid
dinosaurs flourished in what appears to have been a planet with very
y permanent ice and tropical-like fauna and flora existing even at
ide places. Calculations by Barron,[3] now at Penn State University,
s using models with continents put in positions believed to have been
back then suggest that paleocontinental geography alone is not suffi-
plain much more than half of the substantial increase in temperature
) today. Schneider et al., at the National Center for Atmospheric
[4] have suggested from such calculations that it would be easier to

32. Molina, M., and F. S. Rowland. "Stratospheric Sink for Chlorofluoromethanes: Chlorine Atom Catalyzed Destruction of Ozone," *Nature* 249:810 (1974).
33. Remsberg, E.E., J.M. Russell, III, L.L. Gordley, J.C. Gille, and P.L. Bailey. "Implications of the Stratospheric Water Vapor Distribution as Determined by the Nimbus 7 LIMS Experiment," *J. Atmos. Sci.* 41:2934–2945 (1984).
34. Bowman, K. P., and A. J. Krueger. "A Global Climatology of Total Ozone from the Nimbus 7 Total Ozone Mapping Spectrometer," *J. Geophys. Res.* 90:7967–7976 (1985).
35. McPeters, R. D., D. F. Heath, and P. K. Bhartia. "Average Ozone Profiles for 1979 from the Nimbus 7 SBUV Instrument," *J. Geophys. Res.* 89:5199–5214 (1984).
36. Solomon, S. and A. Tuck. "Evaluating Ozone Depletion Potentials," *Nature* 348:203 (1990).
37. Fisher, D. A., M. Ko, D. Wuebbles, and I. Isaksen. "Evaluating Ozone Depletion Potentials," *Nature* 348:203 (1990).

Global Warmin

Stephen H. Schneider**

INTRODUCTION

Hardly anything excites a scientist more than
definitive experiment, especially one capable
theory or settling a hot controversy. As Rev
Institution of Oceanography noted in 1957,[1] w
physical experiment," not in a laboratory, not i
planet. Since the beginnings of the Industrial Re
increased by about 25%, undoubtedly the resul
of this increase is due to the burning of fossil fu
although perhaps 40% is probably due to the d
undertaken in the developed countries in the 1
countries in the latter part of the 20th century
19th century that CO_2 could be an important ele
temperature through what has become known a
a glass in a greenhouse, which traps solar rad
blocking the exchange of warmed inside air w
and particles in the earth's atmosphere preferer
sunlight into the earth's surface than they all
form of radiation to escape to space. This l
atmosphere has been well established by hund
thousands of aircraft and balloon measurem
millions of satellite observations. There is vi
spheric scientists that increasing the concentrat
ane, chlorofluorocarbons, nitrous oxide, and

*An edited version appeared in the September 1989 Sp
Scientific American.

**Any opinions, findings, conclusions, or recommendati
the author and do not necessarily reflect the views of
National Center for Atmospheric Research is sponsore

trace "g
phere a
face. W
will sett
over wh
virtually
scientifi
whether
will incr
in 50, 1
whether
we sho
controv
tial that
slightly,
uncertai

THE CL

Historic

Clima
earth hi
abunda
existenc
tells us t
or below
that the
which p
in solar
such hyp
for any
earth mi
to the M
evidence
15°C wa
down an
little if a
high-lati
and othe
prevalen
cient to
relative
Researcl

reconcile the warm climates of the mid-Cretaceous of 100 million years ago, for example, if there was some other way to heat the planet in addition to altered continental paleogeography. The obvious candidate for extra heating is increased CO_2. Indeed, Robert Berner at Yale University, Anthony Lasaga, then at Penn State, and the late geologist Robert Garrels[5] built a geochemical model suggesting that enhanced rates of sea floor spreading in the Cretaceous times suggested that upwelling magma (and CO_2 along with it) proceeded at a sufficiently fast pace that there should have been much more CO_2 in the air 100 million years ago than at present. They estimate factors of 5 to 10 more CO_2 as plausible, something Russian scientists Ronov and Budyko[6] had speculated on earlier based on the carbonate content of sedimentary rocks. Again, one cannot be quantitative enough about the CO_2 amounts or the earth's temperature change back then to get precise calibrations of what the earth's sensitivity is to given increases of CO_2. However, this piece of historical coincidence is more circumstantial evidence associating an enhanced greenhouse effect with warmer periods on earth.

More recently, exciting geophysical data have been dug out of ice cores in Antarctica. A French glaciological team, headed by Claude Lorius[7] at Grenoble, France, cooperated with a Russian drilling project and examined over 2000 m of ice core taken at Vostok in central Antarctica. This ice at its base was over 140,000 years old, having penetrated 2 ice ages and 2 interglacials (Figure 1). A remarkable finding was uncovered when the ice was brought back to the laboratory at Grenoble in a frozen state and broken down and the trapped gas bubbles in the core analyzed. These gas bubbles were incorporated into the ice at the time the snow was deposited, and as the weight of the snow on top of them compressed that snow into ice, the pore gas became entombed. These air bubbles preserve samples of the atmosphere breathed by our Neanderthal ancestors and the wooly mammoths. What is remarkable in the record is the fact that CO_2 and methane varied over a wide range, mostly together. Even more remarkable is the fact that during the present interglacial, the past 10,000 years, and the previous interglacial, between about 120,000 and 130,000 years ago, the earth's temperature was about 5°C warmer than during the height of the last ice age (about 18,000 years ago) and the ice age before that (about 140,000 years ago). The local temperature change is recorded in the ice by the amount of the heavy isotope ^{18}O (which is relatively absent in cold times). The cores yield a very close correlation between CO_2, methane, and temperature: during the interglacial periods there is about 25% more CO_2 and 100% more methane than during the cold glacial periods. Unfortunately, it is not clear whether this remarkable coincidence is cause and effect (i.e., the CO_2 change caused the ice age) or effect and cause (i.e., the reverse). My own leading speculation is that the ice ages are paced by other factors (such as changes in the orbital elements and the internal dynamical processes associated with glacier buildup and decay time or the response of the bedrock to the glaciers), but that as the ice changes, it induces variations in CO_2 content which substantially amplify the depth of the ice age cold and the warming rate after

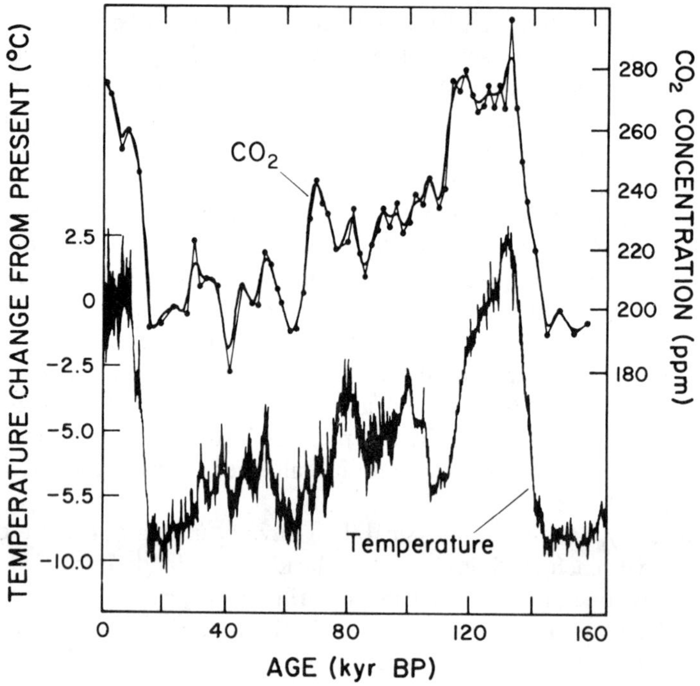

Figure 1. CO_2 concentrations ("best estimates") and smoothed values (spline function) (in ppmv) plotted against age in the Vostok record (upper curves) and local atmospheric temperature change derived from the isotopic profile (lower curve). (Source: Barnola, J. M., D. Raynaud, Y. S. Korotkevich, and C. Lorius. *Nature* 329:408 (1987).)

the ice starts to melt. Although one cannot conclude quantitatively precisely how much temperature change the earth would undergo if CO_2 increased in the future by scaling it to the 25% decrease in the past ice age, the ice core results provide yet another piece of circumstantial evidence suggesting that lower concentrations of greenhouse gases are associated with colder temperatures and increased amounts of greenhouse gases are associated with higher temperatures.

Finally, what has the record of the earth's temperature told us in the past 100 years, a time that has seen about a 25% further increase in CO_2 and nearly another doubling in methane? Two reconstructions of the earth's surface temperature for the past century (Figure 2)* have been made at the Goddard Institute for Space Studies (GISS)[8] and Climatic Research Unit (CRU).[9] Although some identical instrumental records were used in each study, the methods of analysis were different. Moreover, the CRU results include an ocean data set. These records have been criticized because a number of the

*The data from 1989 are about 0.1°C lower than 1988 and preliminary data for 1990 suggest it is the warmest year on record, some 0.05°C warmer on a global basis than the previous record year, 1988.[24]

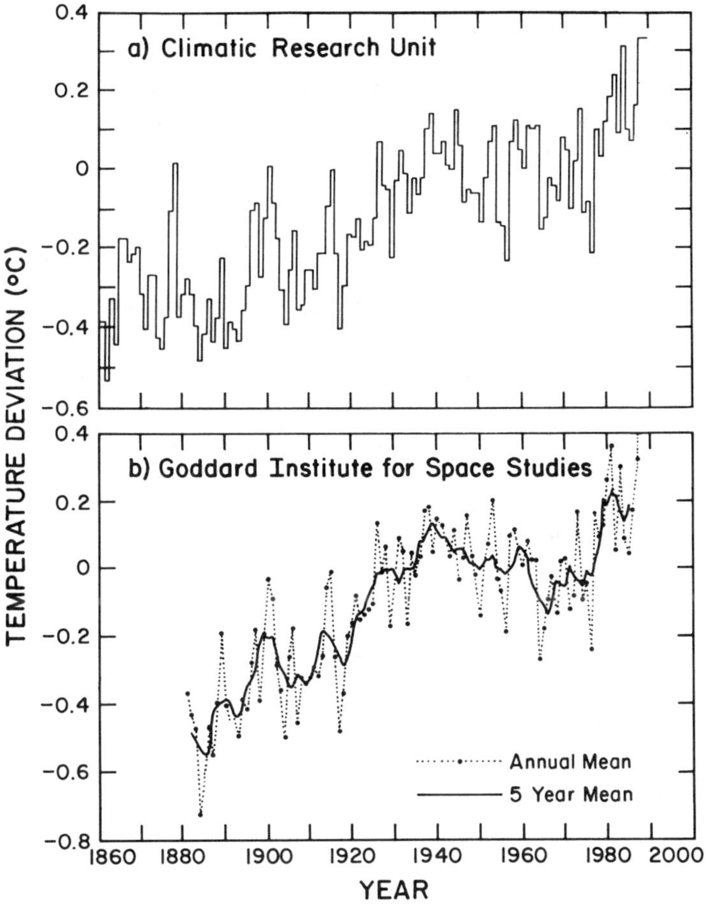

Figure 2. A comparison of the global surface temperature trends of the past 100 years constructed from land and island stations and ocean surface temperature data sets at the (a) Climatic Research Unit and from a similar set of stations (minus the ocean surface temperature data set) at the (b) Goddard Institute for Space Studies. (Source: Jones, P. D. and T. M. L. Wigley, personal communication, and Hansen, J., and S. Lebedeff, *Geophys. Res. Lett.* 15:323, 1988.)

thermometers were in city centers and might have measured a spurious warming from the growth of the urban heat island. In other cases, thermometers were moved from cities to airports or up and down mountains, and some other measurements are also unreliable. A critical evaluation of the urban heat island effect suggests that in the U.S. the data may account for nearly 0.4°C of warming in the GISS record and about 0.15°C warming in the CRU record. Because the U.S. data from where the urban heat island effect might be significant are only a small part of the global total, these corrections should not automatically be made to the entire global record. However, even after such corrections for the U.S. are applied to all of the data, the global data still suggest that nearly 0.5°C warming occurred during the past 100 years.[10] Moreover, the 1980s were the warmest decade on record; 1981, 1987, and 1988 were the warmest years in these records (with preliminary data suggesting 1990 will replace 1988). Once again, there is circumstantial evidence connecting the warming trend with the increased trace gases, but since the warming is not smooth, but rather proceeded in two very rapid jumps with a slight cooling trend in between, it is still difficult to argue cause and effect from this record, as we will explore later.

Scientific Issues Surrounding the Greenhouse Effect

It is helpful to break down the set of issues known as the greenhouse effect into a series of stages, each feeding into another, and then to consider how policy questions might be addressed in the context of these more technical stages.

Projecting Emissions

Behavioral assumptions must be made in order to project future use of fossil fuels (or deforestation, because this too can impact the amount of CO_2 in the atmosphere—it accounts for an additional 20 to 30% of the recent total annual fossil fuel CO_2 injection of about 5.5×10^9 metric tons). The essence of this aspect then is social science. Projections must be made of human population, the per capita consumption of fossil fuel, deforestation rates, reforestation activities, and perhaps even countermeasures to deal with the extra CO_2 in the air. These projections depend on issues such as the likelihood that alternative energy systems or conservation measures will be available, their price, and their social acceptability. Furthermore, trade in fuel carbon (e.g., a large-scale transfer from coal-rich to coal-poor nations) will depend not only on the energy requirements and the available alternatives, but also on the economic health of the potential importing nations. This trade in turn will depend upon whether those nations have adequate capital resources to spend on energy rather than other precious strategic commodities—such as food or fertilizer as well as some other strategic materials such as weaponry. Total CO_2 emissions from energy systems, for example, can be expressed by a formula termed "the population multiplier" by Ehrlich and Holdren[11] nearly two decades ago:

$$\text{Total } CO_2 \text{ emission} = \frac{CO_2 \text{ emission}}{\text{technology}} \times \frac{\text{technology}}{\text{capita}} \times \text{total population size}$$

The first term represents engineering, the second represents standard of living, and the third represents demography in this version, which is modified from the original.

In order to quantify future changes, we can make scenarios that show alternative CO_2 futures based on assumed rates of growth in the use of fossil fuels. Most typical projections are in the 0.5 to 2% annual growth range for fossil fuel use and imply that CO_2 concentrations will double (to 600 ppm) in the 21st century.

Other greenhouse gases like chlorofluorocarbons (CFCs), methane, nitrogen oxides, tropospheric ozone, and others could, together, be as important as CO_2 in augmenting the greenhouse effect, but some of these depend on human behavior and have complicated biogeochemical interactions. These complications account for the large error bars on Figure 3. Furthermore, there are other interesting possible relations between CH_4 and water vapor in the stratosphere, for example, which might affect polar stratospheric clouds which are believed to enhance photochemical destruction of ozone by chlorine atoms. CH_4 also is removed by the atmosphere radical OH^-. Thus, CH_4, by decreasing OH^- can

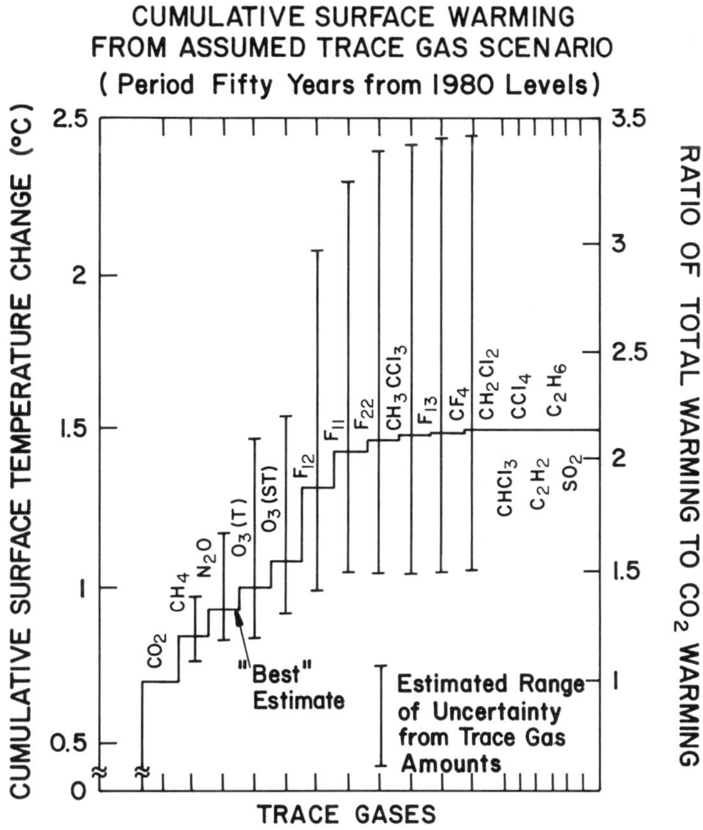

Figure 3. Various trace "greenhouse gases" contribute about as much to equilibrium global warming (see right-hand scale) as CO_2 for the "best" estimate case, but uncertainties in the projected scenarios of these trace gases are large (see vertical bars). Additional uncertainties in equilibrium temperature response from climate model assumptions are not included in the figure. (Source: Ramanathan, V., R. J. Cicerone, H. B. Singh, and J. T. Kiehl. *J. Geophys. Res.* 90:5547 (1985).)

have indirect greenhouse effects by increasing tropospheric ozone concentrations, since low altitude O_3 is removed by OH^-.

Projecting Greenhouse Gas Concentrations

Once a plausible set of scenarios for how much CO_2 will be injected into the atmosphere is obtained, the interacting biogeochemical processes that control the global distribution and stocks of the carbon need to be determined. Such processes involve the uptake of CO_2 by green plants (because CO_2 is the basis of photosynthesis, more CO_2 in the air means faster rates of photosynthesis), changes in the amount of forested area and vegetation type, and how CO_2 fertilization or climate change affects natural ecosystems on land and in the oceans. The transition from ice age to interglacial climates provides a concrete example of how large natural climatic change affected natural ecosystems in North America. This transition represented some 5°C global warming, with as much as 10 to 20°C warming locally near ice sheets. The boreal species now in Canada were hugging the rim of the great Laurentide glacier in the U.S. Northeast some 10,000 years ago, while currently now-abundant hardwood species were restricted to small refuges largely in the South. The natural rate of forest movement that can be inferred is, to order of magnitude, some ~1 km/year, in response to temperature changes averaging ~1 to 2°C/1000 years. If climate were to change much more rapidly than this, then the forests would likely not be in equilibrium with the climate, i.e., they could not keep up with the fast change and would go through a period of transient adjustment in which many hard-to-predict changes in species distribution, productivity, and CO_2 absorptive capacity would likely occur.

Furthermore, because the slow removal of CO_2 from the atmosphere is significantly dependent on biological and chemical processes in the oceans, and because decades to centuries are needed for thermal and dynamical equilibration after a large perturbation, the rates at which climate change modifies mixing processes in the ocean (and thus the oceanic CO_2 uptake rates) also need to be taken into account. There is considerable uncertainty about how much newly injected CO_2 will remain in the air during the next century, but typical estimates put this so-called "airborne fraction" at about 50%. Biological feedbacks can also influence the amount of CO_2 in the air. For example, enhanced photosynthesis could reduce the buildup rate of CO_2 relative to that projected with carbon cycle models that do not include such an effect. On the other hand, although there is about as much carbon stored in the forests as there is in the atmosphere, there is about twice as much carbon stored in the soils in the form of dead organic matter. This carbon is slowly decomposed by soil microbes back to CO_2, CH_4 (in anaerobic soils), and other gases. Because the rate of this decomposition depends on temperature and soil moisture, global warming from increased greenhouse gases could cause enhanced rates of microbial decomposition of necromass (dead organic matter), thereby causing a positive feedback that would enhance CO_2 or CH_4 buildup. Furthermore, considerable methane is trapped below frozen sediments as clathrates in tundra and off continental shelves. These clathrates could release vast quantities of methane into the atmosphere if substantial Arctic warming were to take place. As noted earlier, the ice core data have shown that not only has CO_2 tracked temperature closely for the past 150,000 years, but so has methane, and methane is a significant trace greenhouse gas which is some 20 to 30 times more effective per molecule at absorbing IR radiation than CO_2 (neglecting second-order effects on stratospheric H_2O or tropospheric O_3). Despite these uncertainties, many workers have projected that CO_2 concentrations will reach 600 ppm sometime between 2030 and 2080 and that some of the other trace greenhouse gases will continue to rise at even faster rates.

Estimating Global Climatic Response

Once we have projected how much CO_2 (and other trace greenhouse gases) may be in the air during the next century or so, we have to estimate their climatic effect. Complications arise because of interactive processes, i.e., feedback mechanisms. For example, if added CO_2 were to cause a temperature increase on earth, the warming would likely decrease the regions of the earth covered by snow and ice and decrease the global albedo. The initial warming would thus create a darker planet that would absorb more energy, thereby creating a larger final warming. This scenario is only one of a number of possible feedback mechanisms. Clouds can change in amount, height, or brightness, for example, substantially altering the climatic response to CO_2, and because feedback processes interact in the climatic system, estimating global temperature increases accurately is difficult; projections of the global equilibrium temperature response to an increase of CO_2 from 300 to 600 ppm have ranged from ~1.5 to 5.5°C. (In the next section the much larger uncertainties surrounding regional responses will be discussed.) Despite these uncertainties, there is little debate in scientific circles that continued increases of trace greenhouse gases will cause global warming.

We cannot directly verify our quantitative predictions of greenhouse warming on the basis of purely historical events; therefore, we must base our estimates on numerical climatic models because the complexity of the real world cannot be reproduced in laboratory models. These models are then calibrated against natural analogs of large climatic changes. In the mathematical models, known basic laws are applied to the atmosphere, oceans, land surface, biota, and glaciers, and the equations that represent these laws are solved with the best computers available. Then, we simply change in the computer program the effective amount of greenhouse gases, repeat our calculation, and compare it to the "control" calculation for the present earth. Many such global climatic models have been built during the past few decades, and the results are in rough agreement that if CO_2 were to double from 300 to 600 ppm, then the earth's surface temperature would eventually warm up sometime between 1 and 5°C; the most recent general circulation models' (GCMs) estimates are from 2 to 4°C. For comparison, the global average surface temperature (land and ocean) during the Ice Age extreme 18,000 years ago was only about 5°C colder than that today. Thus, a global temperature change of 1 to 2°C represents a considerable change. A sustained global increase of more than 2°C above present would be unprecedented in the era of human civilization.

The largest uncertainty in estimating the sensitivity of the earth's surface temperature to a given increase in radiative forcing arises from the problem of parameterization. Because the equations that are believed to represent the flows of mass, momentum, and energy in the atmosphere, oceans, land, ice fields, and biosphere cannot be solved analytically with any known techniques, approximation techniques are used in which the equations are discretized with a finite grid that divides the region of interest into cells that are several hundred kilometers or more on a side. Clearly, critically important variables, such as clouds, which largely control the radiation budget of the earth, do not occur on scales as large as the grid of a general circulation model. Therefore, we seek to find a parametric representation or parametrization that relates implicitly the effects of important processes that operate at subgrid scale, but still have effects at the resolution of a typical general circulation model. For example, a parameter or proportionality coefficient might be used that scales the average cloudiness in a grid cell to the mean relative humidity in that cell and some other measures of atmospheric stability. Then, the important task becomes validating these semi-empirical parameterizations because at some scale, all models, no matter how high the resolution, must treat subgrid-scale processes through parameterization.

Projecting Regional Climatic Response

In order to make useful estimates of the effects of climatic changes, we need to determine the regional distribution of climatic change. Will it be drier in Iowa in 2010, too hot in India, wetter in Africa, or more humid in New York; will California be prone to more forest fires or will Miami and Venice flood? Unfortunately, reliable prediction of the time sequence of local and regional responses of variables such as temperature and rainfall requires climatic models of greater complexity and expense than are currently available. Even though the models have been used to estimate the responses of these variables, the detailed regional predictions from state-of-the-art models are not yet reliable.

Although there is considerable experience in examining regional changes, considerable uncertainty remains over the probability that these predicted regional features will occur. The principal reasons for the uncertainty are twofold: the crude treatment in climatic models of biological and hydrological processes and the usual neglect of the effects of the deep oceans. The deep oceans would respond slowly—on time scales of many decades to centuries—to climatic warming at the surface and also act differentially (i.e., nonuniformly in space and through time). Therefore, the oceans, like the forests, would be out of equilibrium with the atmosphere if greenhouse gases increase as rapidly as typically is projected and if climatic warming were to occur as fast as 1 to 6°C during the next century. This typical projection, recall, is 5 to 60 times as fast as the natural average rate of temperature change that occurred from the end of the last Ice Age to the present warm period (i.e., 1 to 6°C warming in a century from human activities compared to an average natural warming of 1 to 2°C per millennium from the waning of the Ice Age to the establishment of the present interglacial epoch). If the oceans are out of equilibrium with the atmosphere, then specific regional forecasts will not have much credibility until fully coupled, three-dimensional, dynamical atmosphere-ocean models are tested and applied. The development of such models is a formidable scientific and computational task and is still not very advanced.

Validation of Climatic Model Forecasts

Of course, it is appropriate to ask how climatic models' predictions of unprecedented climatic change beyond the next several decades might be verified. Can society make trillion dollar decisions about global economic developments based on the projections of these admittedly cloudy crystal balls? How can models be verified?

The first verification method is checking the ability of a model to simulate today's climate. Reproduction of the seasonal cycle is one critical test because these natural temperature changes are several times larger, on a hemispheric average, than the change from an ice age to an interglacial period or a projected greenhouse warming. Also, "fast physics" such as cloud parameterizations can be tested by seasonal simulations or weather forecasts. Global climate models generally map the seasonal cycle of surface temperature very well, which suggests that fast physics can be well simulated on a global basis. However, successful reproduction of these seasonal patterns still won't allow us to claim full validation. Precipitation, relative humidity, and the other variables also need to be checked. Reproduction of the change in daily variance of these variables with the seasons is another tough test. The seasonal tests, however, do not indicate how well a model simulates such medium or slow processes as changes in deep ocean circulation or ice cover, which may have an important effect on the decade to century time scales during which CO_2 or CH_4 concentrations are expected to double.

A second verification technique is isolating individual components of the model, such as a parameterization, and testing them against high-resolution submodels or actual data at high resolution. For example, one can check whether a parameterized evaporation matches the observed evaporation of a particular cell. However, this technique cannot guarantee that the complex interactions of many individual model components are treated properly. The model may predict average cloudiness well, but represent cloud feedback poorly. In this case, simulation of overall climatic response to increased CO_2 is likely to be inaccurate. The seasonal tests or weather forecasts are also needed at high resolution. In addition, a model should reproduce to better than, say, 25% accuracy the flow of thermal energy between the atmosphere, surface, and space. Together, these energy flows comprise the well-established energy balance of earth and constitute a formidable and necessary test for all models. Very recently, Raval and Ramanathan[12] compared model-predicted heat trapping to satellite data of the same quantity with very good results.

A third method for determining overall simulation skill is the model's ability to reproduce past climates or climates of other planets. Paleoclimatic simulations of the Mesozoic Era, glacial-interglacial cycles, or other extreme past climates help in understanding the coevolution of the earth's climate with living things. They are valuable for the estimation of both the climatic and biological future.

Overall validation of climatic models thus depends on constant appraisal and reappraisal of performance in the above categories. Also important are a model's response to such century-long forcings as the 25% increase in CO_2, 100% increase in CH_4 concentrations, and different increases in other trace greenhouse gases since the Industrial Revolution.

Most recent climatic models predict that a warming of at least 1°C should have occurred during the past century. The precise "forecast" of the past 100 years also depends upon how the model accounts for such factors as changes in the solar constant or volcanic dust as well as trace greenhouse gases in addition to CO_2. Indeed, the typical prediction of 1°C warming is broadly consistent, but somewhat larger than that observed. Possible explanations for the discrepancy include

1. The state-of-the-art models are too sensitive to increases in trace greenhouse gases by a rough factor of 2.
2. Modelers have not properly accounted for such competitive external forcings as volcanic dust or changes in solar energy output.
3. Modelers have not accounted for other external forcings such as regional tropospheric aerosols from agricultural, biological, and industrial activity.
4. Modelers have not properly accounted for internal processes that could lead to stochastic or chaotic behavior.
5. Modelers have not properly accounted for the large heat capacity of the oceans taking up some of the heating of the greenhouse effect and delaying, but not ultimately reducing, warming of the lower atmosphere.
6. Both present models and observed climatic trends could be correct, but models are typically run for equivalent doubling of CO_2 concentration, whereas the world has only experienced a quarter of this increase and nonlinear processes have been properly modeled and produced a sensitivity appropriate for doubling, but not for 25% increase.
7. The incomplete and inhomogenous network of thermometers has underestimated actual global warming this century.

Despite this array of excuses why observed global temperature trends in the past century and those anticipated by most GCMs disagree somewhat, the roughly twofold discrepancy between predicted and measured temperature changes is not large, but still of concern. This rough validation is reinforced by the good simulation by most climatic models of the seasonal cycle of surface temperature or monsoon rainfall, diverse ancient paleoclimates, hot conditions on Venus, cold conditions on Mars (both well simulated), and the present distribution of climates and radiation fluxes on earth. When taken together, these verifications provide strong circumstantial evidence that the current modeling of the sensitivity of global surface temperature to given increases in greenhouse gases over the next 50 years or so is probably valid within a rough factor of two. Most climatologists do not yet proclaim that the observed temperature changes this century were caused beyond doubt by the greenhouse effect. The relation between the observed century-long trend and the predicted warming could still be chance occurrences, or other factors, such as solar constant variations or volcanic dust or industrial emission of sulfur dioxide, may not have been accounted for correctly during the past century—except during the past decade when accurate measurements began to be made.

Another decade or two of observations of trends in earth's climate, of course, should produce signal-to-noise ratios sufficiently high that we will be able to determine conclusively the validity of present estimates of climatic sensitivity to increasing trace greenhouse gases. That, however, is not a cost-free enterprise because a greater amount of change could occur then than if actions were undertaken now to slow down the buildup rate of greenhouse gases.[13]

Scenarios of the Environmental Impact of CO_2

Given a set of scenarios for regional climatic change, we must next estimate the impacts on the environment and society. Most researchers have focused on the direct effects of CO_2 increases or use model-predicted maps of temperature and rainfall patterns to estimate impacts on crop yields or water supplies. Also of concern is the potential that temperature increases will alter the range or numbers of pests that affect plants or diseases that threaten animals or human health. Also of interest are the effects on unmanaged ecosystems, principally forests. For example, ecologists are concerned that the destruction rate of tropical forests attributed to human expansion is eroding the genetic diversity of the planet, i.e., because the tropical forests are in a sense major banks for the bulk of living genetic materials on earth, the world is losing some of its irreplaceable biological resources through rapid development. Substantial changes in tropical rainfall have been predicted on the basis of climatic models; reserves (or refugia) that are currently set aside as minimal solutions for the preservation of some genetic resources into the future may not even be as effective as currently planned.[14]

Climate changes resulting from greenhouse gas increases could also significantly affect water supply and demand. For example, a local increase in temperature of several degrees Celsius could decrease runoff in the Colorado

River Basin by tens of percent. A study by Gleick[15] of the vulnerability to climate change of various water resource regions in the U.S. showed that some regions are quite vulnerable to climatic changes (Table 1).

Water quality will be diminished if the same volume of wastes are discharged through decreased stream flow. In addition, irrigation demand (and thus pressure on ground-water supplies) may increase substantially if temperatures increase without concomitant offsetting increases in precipitation. A number of climate models suggest that temperatures could increase and precipitation decrease simultaneously in several areas, including the central plains of the U.S. Peterson and Keller[16] estimated the effects of a 3°C warming and a 10% precipitation change on U.S. crop production based on crop water needs. The greatest impact would be in the western states and the Great Plains, with less in the northwest. The warm, dry combination would increase depletion of streams and reduce viable acreage by nearly a third in the arid regions. New supplies of water would be needed, threatening ground water and the viability

Table 1. The Vulnerability to Climate Change of the Water-Resource Regions of the U.S.

Vulnerability Category: Basin	Lamps On	Consumption (D/Q)	Storage (Q/S)	Variability (Q05/Q95)	Hydro-electric (H/E)	Ground Water (GO/GW)
Great Basin	5	*	*	*	*	*
Missouri	4	*		*	*	*
California	4	*	*	*	*	
Arkansas–White-Red	3		*	*		*
Texas-Gulf	3	*		*		*
Rio Grande	3	*		*		*
Lower Colorado	3	*			*	*
Tennessee	2		*		*	
Lower Mississippi	2		*	*		
Upper Colorado	2	*		*		
Pacific Northwest	2		*		*	
Alaska	2		*		*	
Caribbean	2		*	*		
New England	1		*			
Mid-Atlantic	1		*			
South Atlantic-Gulf	1		*			
Great Lakes	1		*			
Ohio	1		*			
Upper Mississippi	1		*			
Souris-Red-Rainy	1			*		
Hawaii	1		*			

Note: The vulnerability to climate change at water resource regions is shown by ratios defined below: D/Q for high consumption, Q/S for little storage, Q05/Q95 for high variability, H/E for dependent on hydroelectricity, and GO/GW for Groundwater overdraft. If the ratio exceeds a threshold defined in (Reference 15), the warning lamp is "on" as indicated by an asterisk (*). The column regions with most "on" are at the top of the table. The table suggests which U.S. basins would be most jeopardized by increases in temperature that decrease runoff by increasing evaporation or those which could benefit most should the climate change increase precipitation sufficiently to overcome any increases in evaporation caused by temperature increase.

of agriculture in these regions. On the other hand, farmers in the East, and particularly in the Southeast, might profit if the depletion of eastern rivers were relatively less severe than that in the West or the Great Plains. However, increases in the efficiency of the irrigation management and technological improvements remain achieveable and would help substantially to mitigate potential negative effects. Drying in the West could also markedly increase the incidence of wildfires, which in turn could act as agents of rapid ecological change.

Most workers project that an increase in global temperature of several degrees Celsius could cause sea level to rise by up to 1.0 m generally in the next 50 to 100 years; such a rise would endanger coastal settlements, estuarine ecosystems, and the quality of coastal freshwater resources. However, there are many uncertainties and this very important impact area remains quantitatively unclear.[17]

Economic, Social, and Political Impacts

The estimation of the distribution of economic "winners and losers", given a scenario of climatic change, involves more than simply looking at the total dollars lost and gained—were it possible somehow to make such a calculation credibly! It also requires looking at these important equity questions: "who wins and who loses?" and "how might the losers be compensated and the winners charged?" For example, if the cornbelt in the U.S. were to "move" north and east by several hundred kilometers from a warming, then a billion dollars a year lost in Iowa farms could well eventually become Minnesota's billion dollar gain. Although some macroeconomists viewing this hypothetical problem from the perspective of the U.S. as a whole might see no net losses here, considerable social consternation could be generated by such a shift in climatic resources, particularly since the cause was economic activities (e.g., CO_2 production) that directed differential costs and benefits to various groups. Moreover, even the perception that the economic activities of one nation could create climatic changes that would be detrimental to another has the potential for disrupting international relations—as is already occurring in the case of acid rain. In essence, what greenhouse gas-induced environmental changes create is an issue of "redistributive justice."

If a summer soil moisture decrease, such as projected for the U.S. by some GCMs were to occur, then it would have disturbing implications for agriculture in the U.S. and Canadian plains. Clearly, present farming practices and cropping patterns would have to change. The more rapidly the climate changed and the less accurately the changes were predicted (which go together), the more likely that the net changes would be detrimental. It has been suggested that a future with summer soil moisture decrease could translate to a loss of comparative advantage of U.S. agricultural products on the world market. Such a scenario could have substantial economic and security implications. Taken together, projected climate changes into the next century could have major impacts on water resources, sea level, agriculture, forests, biological diversity, air quality, human health, urban infrastructure, and electricity demand.

Policy Responses

The last stage in diagnosing the greenhouse effect concerns the question of appropriate policy responses. Three classes of actions could be considered. First, engineering countermeasures: purposeful interventions in the environment to minimize the potential effects (e.g., deliberately spreading dust in the stratosphere to reflect some extra sunlight to cool the climate as a countermeasure to the inadvertent CO_2 warming). These "geoengineering" countermeasures suffer from the immediate and obvious flaw that if there is admitted uncertainty associated with predicting the unintentional consequences of human activities, then likewise substantial uncertainty surrounds any deliberate climatic modification. Thus, it is quite possible that the unintentional change might be overestimated by computer models and the intentional change underestimated, in which case human intervention would be a "cure worse than the disease." Furthermore, the prospect for international tensions resulting from any deliberate environmental modifications is staggering, and our legal instruments to deal with these tensions are immature. Thus, acceptance of any substantial climate countermeasure strategies for the foreseeable future is hard to imagine, particularly because there are other more viable alternatives.[18]

The second class of policy action, one that tends to be favored by many economists, is adaptation.[19] Adaptive strategists propose to let society adjust to environmental changes. In extreme form, some believe in adaptation without attempting to mitigate or prevent the changes in advance. Such a strategy is based partly on the argument that society will be able to replace much of its infrastructure before major climatic changes materialize and that because of the large uncertainties, we are better off waiting to see what will happen before making potentially unnecessary investments. However, it appears quite likely that we are already committed to some climatic change based on emissions to date, and therefore some anticipatory steps to make adaptation easier certainly seem prudent. We could adapt to climate change, for example, by planting alternative crop strains that would be more widely adapted to a whole range of plausible climatic futures. Of course, if we don't know what is coming or we haven't developed or tested the seeds yet, we may well suffer substantial losses during the transition to the new climate. However, such adaptations are often recommended because of the uncertain nature of the specific redistributive character of future climatic change and because of high discount rates that devalue future risks.

In the case of water supply management, an American Association for the Advancement of Science Panel on Climate Change[20] made a strong, potentially controversial, but, I believe, rather obvious, adaptive suggestion: governments at all levels should reevaluate the legal, technical, and economic components of water supply management to account for the likelihood of climate change, stressing efficient techniques for water use and new management practices to increase the flexibility of water systems and recognizing the need to reconsider existing compacts, ownership, and other legal baggage associated with the present water system. In light of rapid climate change, we need to reexamine the balance between private rights and the public good, because water is intimately connected with both. This may be politically sensitive, but ideology is the enemy of flexibility, and flexibility is the best hedge against the prospect of substantial changes. Regional transfers from water-abundant regions to water-deficient regions are often prohibited by legal or economic impediments that need to be examined as part of a hedging strategy for adapting more effectively to the prospect of climatic change even though regional details cannot now be reliably forecast.

Finally, the most active policy category is prevention, which could take the form of sulfur scrubbers in the case of acid rain, abandonment of the use of chlorofluorocarbons and other potential ozone-reducing gases (particularly those that also enhance global warming), and reduction in the amount of fossil fuel used around the world or fossil fuel switching from more CO_2- and SO_2-producing coal to cleaner, less polluting methane fuels. Prevention policies, often advocated by environmentalists, are controversial because they involve, in some cases, substantial immediate investments as insurance against the possibility of large future environmental change, change whose details cannot be predicted precisely. The sorts of preventive policies that could be considered are increasing the efficiency of energy production and end use, the development of alternative energy systems that are not fossil-fuel based, or, in a far-reaching proposal: a "law of the air" proposed in 1976 by Mead and Kellogg.[21] They suggest that various nations would be assigned polluting rights to keep CO_2 emissions below some agreed global standard. A "law of the atmosphere" was recently endorsed in the report of a major international meeting.[22]

HIGH LEVERAGE ACTIONS TO COPE WITH GLOBAL WARMING

Clearly, society does not have the resources to hedge against all possible negative future outcomes. Is there, then, some simple principle that can help us choose which actions to spend our resources on? One guideline is called the "tie-in strategy". Quite simply, society should pursue those actions which provide widely agreed societal benefits even if it is not certain that the predicted change will materialize. For instance, one of the principal ways to slow down the rate at which the greenhouse effect will be enhanced is to invest in more efficient use and production of energy. More efficiency, therefore, would reduce the growing disequilibrium among physical, biological, and social systems and could buy time both to study the detailed implications of the greenhouse effect further and to ensure an easier adaptation. However, if the greenhouse effects now projected prove to be substantial overestimates, what would be wasted by an energy efficiency strategy? Efficiency usually makes good economic sense (although the rate of investment in efficiency does depend, of course, on other competing uses of those financial resources and on the discount rate used). However, reductions in emissions of fossil fuels, especially coal, will certainly reduce acid rain, limit negative health effects in crowded areas from air pollution, and lower dependence on foreign sources of fuel, especially oil, and they will reduce the balance of payments deficits in countries like the U.S. In addition, more energy-efficient factories mean reduced energy costs for manufacturing and thus greater long-term product competitiveness against foreign producers.

Development of alternative, environmentally safer energy technologies (possibly including methane or hydrogen fuels, solar photovoltaic collectors, passively safe nuclear power, or some form of fusion) is another example of a tie-in strategy, as is the development and testing of alternative crop strains, trading agreements with nations for food or other climatically dependent strategic commodities, and so forth. However, there is some ideological opposition to such strategies on the grounds that these activities should be pursued by individual investment decisions through a market economy, not by collective action using tax revenues or other incentives. In rebuttal, a market which does not include the costs of environmental disruptions can hardly be considered a truly free market. Furthermore, strategic investments are made routinely on noneconomic (i.e., cost-benefit analyses are secondary) criteria even by the most politically conservative people: to purchase military security. A strategic consciousness, not an economic calculus, dictates the scale of investments in defense. Similarly, people purchase insurance as a hedge against plausible, but uncertain, future problems. The judgment here is whether strategic consciousness, widely accepted across the political spectrum, needs to be extended to other potential threats to security, including a substantially altered environment occurring on a global scale at unprecedented rates. Then, the next problem is to determine how many resources to allocate.

If we choose to wait for more scientific certainty over details before preventive actions are initiated, then this is done at the risk of our having to adapt to a larger, faster occurring dose of greenhouse gases than if actions were initiated today. In my value system, high-leverage, tie-in actions are long overdue. Of course, I repeat that whether to act is not a scientific judgment, but a value-laden political choice that cannot be resolved by scientific methods.

Incentives for investments to improve energy efficiency, to develop less polluting alternatives, to control methane emissions, or to phase out CFCs may require policies that charge user fees on activities in proportion to the amount of pollution each generates. This strategy might differentially impact less developed nations or segments of the population such as coal miners or the poor. Indeed, an equity problem is raised through such strategies. Finding a balance between fairness and effectiveness is essential. However, is it more appropriate to subsidize poverty, for example, through artificially lower prices of energy which ignore environmental costs, distort the market, and discourage efficient energy end use or alternative production, or is it better to fight poverty by direct economic aid? Perhaps, fairness suggests targeting some fraction of a carbon-energy tax to help those immediately disadvantaged. In addition, this would improve the political tractability of any attempt to internalize the external costs of pollution not currently charged to energy production or end use.[23] In any case, consideration of these political issues will be essential if global-scale agreements are to be negotiated, and without global-scale agreements, no nation acting alone can reduce global warming by more than 10% or so.

A SCIENTIFIC CONSENSUS?

In summary, a substantial warming of the climate through the augmentation to the greenhouse effect is very likely if current technological, economic, and demographic trends continue. Rapid climatic changes will cause both ecological and physical systems to go out of equilibrium—a transient condition that makes detailed predictions tenuous. The faster the changes take place, the more difficult it is to forecast the detailed changes reliably and the less societies or natural ecosystems will be able to adapt to them without potentially serious disruptions. Both the rate and magnitude of typical projections up to 2050 suggest climatic changes beyond that experienced by civilization could occur. The faster the climate is forced to change, the more likely there will be unexpected surprises lurking. The consensus about the likelihood of future global change weakens over detailed assessments of the precise timing and geographic distribution of potential effects and crumbles over the value question of whether present information is sufficient to generate a societal response stronger than more scientific research on the problems—appropriate (but self-serving) advice which we scientists, myself included, somehow always manage to recommend.

I am frequently asked if I am pessimistic because of the fact that it is impossible to avert some global change; it appears that we are already committed to a degree or two of warming, virtually no matter what plausible policies are undertaken. Actually, I view the prospect of global warming not entirely negatively. The potential threat that the larger scenarios pose may be sufficient to catalyze international actions to make environmentally sustainable development a number one world priority. Under those conditions, although we may not avert some change, we could substantially lower the final amount of global change and, perhaps more importantly, stretch it out in time. To me, it would be worth some risk of temporarily allowing global change if we were able to stabilize population growth and provide energy-efficient and environmentally safe technologies sooner as a result of global cooperation to limit the ultimate amount of greenhouse gases that would otherwise be injected by a growing world population that uses energy and materials without consideration of emissions.

The resources that may be needed for this transition to a stabilized, sustainable world are probably in the range of tens of billions of dollars a year for many decades. While it is easy to be pessimistic that such resources will be mobilized, not many years ago it would have seemed inconceivable that massive disengagement of NATO and Warsaw Pact forces in Europe was possible in the foreseeable future. It now seems to me that such disengagement is more than possible, but even likely before the end of this century. The reduced need for hundreds of billions of dollars to be spent on defense budgets on both sides of the East/West gap could free the kind of resources needed both to improve the economic base of those countries as well as provide technology and resources needed by developing countries. We need to ensure, for example, that the energy production and use infrastructure that countries such as China and India will lock into place for the next 50 years will be the highest-efficiency technologies, rather than initially cheaper but more inefficient, highly polluting, and in the long run, more expensive technologies. We have a historic opportunity now, where developing countries can choose between the wasteful and highly polluting systems that developed countries indulged in in the earlier part of the 20th century or, rather, take advantage of the technological and social gains of the last several decades and invest in both economically more viable and environmentally more sustainable technological systems. Unfortunately, this choice will take a combination of more upfront capital, substantial technology transfer, and a much larger level of cooperation and trust than has been heretofore possible. Let's hope that the threat of rapidly growing global changes might be just the impetus the world needs to act in concert to protect the planet against the very real possibility of dangerous alterations to our environmental systems.

REFERENCES

1. Revelle, R., and H. Suess. "Carbon Dioxide Exchange Between Atmosphere and Ocean and the Question of an Increase of Atmospheric CO_2 During Past and Present Decades," *Tellus* 9:18–27 (1957).
2. Ramanathan, V., R. J. Cicerone, H. B. Singh, and J. T. Kiehl. "Trace Gas Trends and Their Potential Role in Climate Change," *J. Geophys. Res.* 90:5547–5566 (1985).
3. Barron, E. J. "Climate Models for the Pre-Pleistocene," in *Paleoclimate Analysis and Modeling*, A. Hecht, Ed. (New York: John Wiley & Sons, Inc., 1985), pp. 397–421.
4. Schneider, S. H., S. L. Thompson, and E. J. Barron. "Mid-Cretaceous, Continental Surface Temperatures: Are High CO_2 Concentrations Needed to Simulate Above-Freezing Winter Conditions?," in *The Carbon Cycle and Atmospheric CO_2: Natural Variations Archean to Present*, E. Sundquist and W. Broecker, Eds.

(Washington, D.C.: Geophysical Monograph Series, Vol. 32, American Geophysical Union, 1985), pp. 554-559.

5. Berner, R. A., A. C. Lasaga, and R. M. Garrels. "The Carbonate-Silicate Geochemical Cycle and Its Effect on Atmospheric Carbon Dioxide over the Past 100 Million Years," *Am. J. Sci.* 283:641–683 (1988).
6. Budyko, M. I., and A. B. Ronov. "Chemical Evolution of the Atmosphere in the Phanerozoic," *Geochem. Int. 1979* 16:1-9 (1980).
7. Barnola, J. M., D. Raynaud, Y. S. Korotkevich, and C. Lorius. "Vostok Ice Core Provides 160,000-Year Record of Atmospheric CO_2," *Nature* 329:408–414 (1987).
8. Hansen, J., and S. Lebedeff. "Global Trends of Measured Surface Air Temperature," *J. Geophys. Res.* 92:13,345–13,372 (1987).
9. Jones, P. D., and T. M. L. Wigley. Private communication (updating a long series of published works) (1989).
10. Karl, T. R., and P. D. Jones. "Urban Bias in Area-averaged Surface Air Temperature Trends," *Bull. Am. Meteorolog. Soc.* 70:265–270 (1989).
11. Ehrlich, P. R., and J. P. Holdren. "Impact of Population Growth," *Science* 171:1212–1217 (1971).
12. Raval, A., and V. Ramanathan. "Observational Determination of the Greenhouse Effect," *Nature* 342:758 (1989).
13. Schneider, S. H. *Global Warming: Are We Entering the Greenhouse Century?* (San Francisco: Sierra Club Press, 1989), 317 pp.
14. Wilson, E. O. "Threats to Biodiversity," *Sci. Am.* 261(3):108–116 (1989).
15. Gleick, P. "Vulnerabilities of Water Systems," in *Climate Change and U.S. Water Resources*, P. E. Waggoner, Ed. (New York: John Wiley & Sons, Inc., 1990), pp. 223–240.
16. Peterson, D. F., and A. A. Keller. "Irrigation," in *Climate Change and U.S. Water Resources*, P. E. Waggoner, Ed. (New York: John Wiley & Sons, Inc., 1990), pp. 269–306.
17. Warrick, R. A., and H. Oerlemans. "Sea Level Rise," in *Climate Change: The IPCC Scientific Assessment*, J. T. Houghton, G. J. Jenkins, and J. J. Ephraums, Eds. (Cambridge: Cambridge University Press, 1990), pp. 257–281.
18. Kellogg, W. W., and S. H. Schneider. "Climate Stabilization: For Better or Worse?," *Science* 186:1163–1172 (1974).
19. Schelling, T. C. "Climatic Change: Implications For Welfare and Policy," in *Changing Climate*, Report of the Carbon Dioxide Assessment Committee (Washington, D.C.: National Academy Press, 1983), pp. 449–482.
20. Waggoner, P. E., Ed. *Climate Change and U.S. Water Resources* (New York: John Wiley & Sons, Inc., 1990), p. 496.
21. Kellogg, W. W., and M. Mead, Eds. *The Atmosphere: Endangered and Endangering*, Fogerty International Center Proceedings No. 39 (Washington, D.C.: U.S. Government Printing Office, 1974), p. 154.
22. Conference Statement. The Changing Atmosphere: Implications for Global Security, Toronto, Ontario, Canada, June 27 to 30, 1988, Environment Canada (Toronto, 1988).
23. Schneider, S. H. "Coping with the Greenhouse Century" in *Global Warming: Are We Entering the Greenhouse Century?* (San Francisco: Sierra Club Books, 1989), pp. 238–285.
24. Jones, P. D. Personal communication.

CHAPTER 4

Interaction of Atmospheric Chemical and Climate Change: Implications for Tropospheric Ozone

Anne M. Thompson

INTRODUCTION

Tropospheric ozone (O_3) is one of the atmosphere's oxidants, along with OH (the hydroxyl radical) and hydrogen peroxide (H_2O_2), the oxidant that acts in clouds and rainwater to convert SO_2 to sulfuric acid. These oxidants are all subject to perturbation by increases in atmospheric trace gases (e.g., methane and CO) as well as changes that might occur over the next few decades in stratospheric ozone and climate. Indeed, tropospheric ozone is on the increase in at least some nonurban northern mid-latitude locations.[1-3] This is of concern because tropospheric ozone is a greenhouse gas[4] as well as an indicator of air quality.

We have completed several studies[5-10] aimed at determining how continuing increases in methane (CH_4) which has doubled in the atmosphere since 1700),[11-13] carbon monoxide (CO, currently increasing 1%/year),[14] and nitrogen oxides (from NO, a pollutant from combustion) have perturbed tropospheric O_3 in the past and are likely to change O_3 over the next few decades. We have also taken a preliminary look at how an increase in ultraviolet (UV) (from stratospheric O_3 depletion) and an increase in tropospheric water (one possible consequence of a global warming)[15] could interact with ozone in the troposphere.

We have done this by calculating O_3 (and OH and H_2O_2) changes from 1985 to 2035 using a simple one-dimensional model to simulate prototype regions that are characterized by a certain "chemical coherence." These include northern mid-latitude urban areas, nonpolluted continental and marine environments, and continental and marine tropical regions. Global changes in O_3, OH, and H_2O_2 are estimated by averaging over the individual regions.

METHOD

Chemically Coherent Regions

The term "chemically coherent" is used to distinguish regions of various mean chemical composition. The photochemical model is initialized using boundary conditions that simulate six chemically coherent regions because there is a strong sensitivity of OH and O_3 changes to background NO and hydrocarbon levels.[16,17] We use a recently published global NO emissions inventory[18] to divide northern mid-latitudes into three regions: "Urban", "Clean Continental", and "Marine". Urban regions do not simulate highly urbanized regions, but have O_3 and NO_x ($NO_x = NO + NO_2$) at levels found in large regions that are urban-influenced:[19] O_3 = 50 to 60 ppbv at the surface for present-day conditions; NO_x = 1 to 2 ppbv. The region that is designated "Clean Continental" refers to nonpolluted continental regions with NO_x levels 10 to 15% as great as Urban regions; O_3 is 40 to 45 ppbv at present. "Marine" regions have NO_x levels at only 20 to 30 pptv and O_3 is 24 ppbv.

There are two low-latitude (tropical) regions, one Continental and one Marine. The sixth region is assigned to southern hemisphere mid-latitudes and resembles the marine northern hemisphere region in NO_x and O_3 levels. CO levels are about half of Northern Hemisphere concentrations,[20,21] however.

Table 1 summarizes surface mixing ratios of O_3, NO_x, CO, and H_2O_2 for the chemically coherent regions we simulate. These six regions cover latitudes between 63N and 63S and represent 90% of the earth's surface.

Photochemical Model Calculations of Perturbed Scenarios

The photochemical model[6] used for calculation of perturbed oxidant levels is a standard one-dimensional transport-kinetics model in which vertical transport is represented by eddy diffusion and 73 photochemical reactions are used to formulate continuity equations for 24 species: O_3, $O(^3P)$, H, HO_2, OH, H_2O_2, NO, NO_2, NO_3, N_2O_5, HNO_3, HNO_4, CH_3, CH_3O_2, CH_3OOH, CH_3O, CH_2O, CO, C_2H_6, $C_2H_5O_2$, C_2H_5OOH, CH_3CHO, CH_3CO_3, and PAN. Vertical profiles of CH_4, H_2O, and H_2 are specified; for CH_4 the values at the ground are based on observations. CO and NO emissions are simulated by a flux at the model lower boundary (0 km). O_3 has an upper boundary (15 km) influx from the stratosphere and is removed at the lower boundary by dry deposition with a rate that is region-dependent according to the predominant surface.[22] (e.g., 0.4 cm s^{-1} for Urban, 0.2 cm s^{-1} for Continental, 0.1 cm s^{-1} for Marine). The transient OH is transported in the model continuity equation, but is essentially in photochemical equilibrium. H_2O_2 is removed at the surface and is also removed by rainout in the lower troposphere with a first-order rate coefficient that is equivalent to a 1-week removal time.

Perturbation calculations are made assuming that the starting year is 1985 with "unperturbed" conditions as described in Table 1. A scenario consists of a

Table 1. Composition of Unperturbed Chemically Coherent Regions (Taken for 1985)

Region	% Area (63N–63S)	Mixing Ratios				Number Density
		NO_x (ppbv, 0 km)	CO (ppbv, 0 km)	O_3 (ppbv, 0 km)	H_2O_2 (ppbv, 0 km)	OH (cm^{-3}, 0 km)
Urban mid-latitude (Urban)	3.7	1.4	362	59	0.12	6.4 (5)*
Clean Continental Mid-latitude (Continental)	9.7	0.20	134	42	0.31	1.1 (6)
Marine mid-latitude (Marine)	7.7	0.020	110	24	0.22	3.0 (5)
Marine low-latitude (15N, Marine low latitude)	32.9	0.023	86	17	0.61	1.1 (6)
Continental low-latitude (15N, Continental Tropical)	24.9	0.073	88	22	0.82	1.8 (6)
Southern Hemisphere Mid-Latitude (SHML)	21.1	0.021	66	25	0.22	5.2 (5)

Note: 6.4 (5) followed by asterisk (*) signifies 6.4×10^5.

series of steady-state model runs, each simulating perturbed conditions at 5-year intervals between 1985 and 2035. At each 5-year point, altered conditions are simulated by varying specified CH_4 mixing ratios and fluxes of CO and NO. The validity of using the steady-state model in this manner was verified by running a time-dependent version of the model for selected scenarios. Changes in UV due to perturbed stratospheric ozone are represented by reducing the above troposphere O_3 column for computation of photodissociation rates. The influx of O_3 from stratosphere to upper troposphere is not changed in the perturbation calculation because stratospheric chemical models indicate that ozone depletion is strongest in the mid stratosphere and is nearly unchanged in the lower stratosphere from which intrusions into the troposphere usually occur. The perturbation to climate is simulated by altering the specified vapor profile in accordance with a 2 K increase in temperature, assuming that relative humidity is constant. For a given scenario, the perturbations are run for all chemically coherent regions.

Three scenarios were considered (Table 2). In the first scenario ("chemistry only"), it is assumed that CH_4 and CO increase at roughly current rates (0.5 to 1%/year) in all regions. No change in NO emissions is assumed because these emissions are restricted to small areas. This globally uniform scenario permits comparison of large-scale trends on the background chemistry of the various regions. The second scenario, "chemistry + UV", is used only for 2035 and assumes that the CH_4 and CO increases of the first scenario occur against a background of decreasing stratospheric O_3 (10% loss from 1985 to 2035). More UV will be transmitted to the troposphere in such a case. The third scenario, "chemistry + UV + more H_2O", assumes that in 2035 there will be more water vapor in the troposphere as well as more UV, CH_4, and CO.

RESULTS

Figure 1 shows tropospheric O_3, OH, and H_2O_2 changes at 2035 (total column, 0 to 15 km, normalized to 1985) for the chemistry-only scenarios for the six regions. Ozone increases in response to CH_4 and CO increases, due to the following reactions:

$$CO + OH\ (+\ O_2) \rightarrow CO_2 + HO_2 \tag{1}$$

$$HO_2 + NO \rightarrow NO_2 + OH \tag{2}$$

or

$$CH_4 + OH\ (+O_2) \rightarrow CH_3O_2 + H_2O \tag{3}$$

$$CH_3O_2 + NO \rightarrow NO_2 + CH_3O \tag{4}$$

The free radical is further oxidized to formaldehyde (H_2CO) and CO. The photolysis of NO_2 leads to ozone formation:

Table 2. Scenarios for 50–Year Runs: 1985 to 2035

Scenario	CH_4*	CO	NO	Stratospheric O_3	H_2O
1 Chemistry only	+ 0.8%/year mix. rat.[a]	+ 0.5%/year flux	Constant	Constant	Constant
2 Chemistry + UV	+ 0.8%/year mix. rat.	+ 0.5%/year flux	Constant	−10% at 50 years	Constant
3 Chemistry + UV + H_2O	+ 0.8%/year mix. rat.	+ 0.5%/year flux	Constant	−10% at 50 years	+ 12% at 50 years

Note: Asterisk (*) indicates mixing ratios for CH_4 (ppmv);

	1985	1995	2005	2015	2025	2035
Northern Hemisphere	1.663	1.815	1.963	2.115	2.265	2.375
Southern Hemisphere	1.537	1.664	1.802	1.951	2.113	2.289

[a]mix. rat. = mixing ratios.

$$NO_2 + h\upsilon \rightarrow NO + O \quad (5)$$

$$O + O_2 + M \rightarrow O_3 + M \quad (6)$$

Figure 1a shows that the greatest increase in O_3 occurs in the Urban region, where NO levels are highest and Reactions 2 and 4 are efficient.

Figure 1b shows OH changes as CO and CH_4 increase from 1985 to 2035. As Reactions 1 and 3 suggest, OH decreases as CO and CH_4 increase and there is a 50-year loss in OH in all regions. However, the losses are greatest in lower NO_x regions (Marine mid-latitude and Southern Hemisphere mid-latitude) and only 5% in the Urban region. This is because O_3 is the source of OH:

$$O_3 + h\upsilon \rightarrow O_2 + O(^1D) \quad (7)$$

$$O(^1D) + H_2O \rightarrow OH + OH \quad (8)$$

and in the urban regions, where O_3 increases most sharply from 1985 to 2035, additional OH forms competitively with the loss imposed by increasing CH_4 and CO (Reactions 1 and 3). In the more remote (lower NO_x) regions increased consumption by CH_4 and CO dominates and OH losses from 1985 to 2035 are 10 to 15%.

The fractional changes in H_2O_2 between 1985 and 2035 are greater than those for O_3 and OH (Figure 1c). Two effects are responsible. The first is the increase in O_3 in all regions, producing more HO_x (OH + HO_2) from Reactions 1, 7, and 8. The second effect is conversion of OH to HO_2 through Reactions 1 and 3, raising the $[HO_2]/[OH]$ ratio. Two HO_2 radicals combine to form:

$$HO_2 + HO_2 \rightarrow H_2O_2 + O_2$$

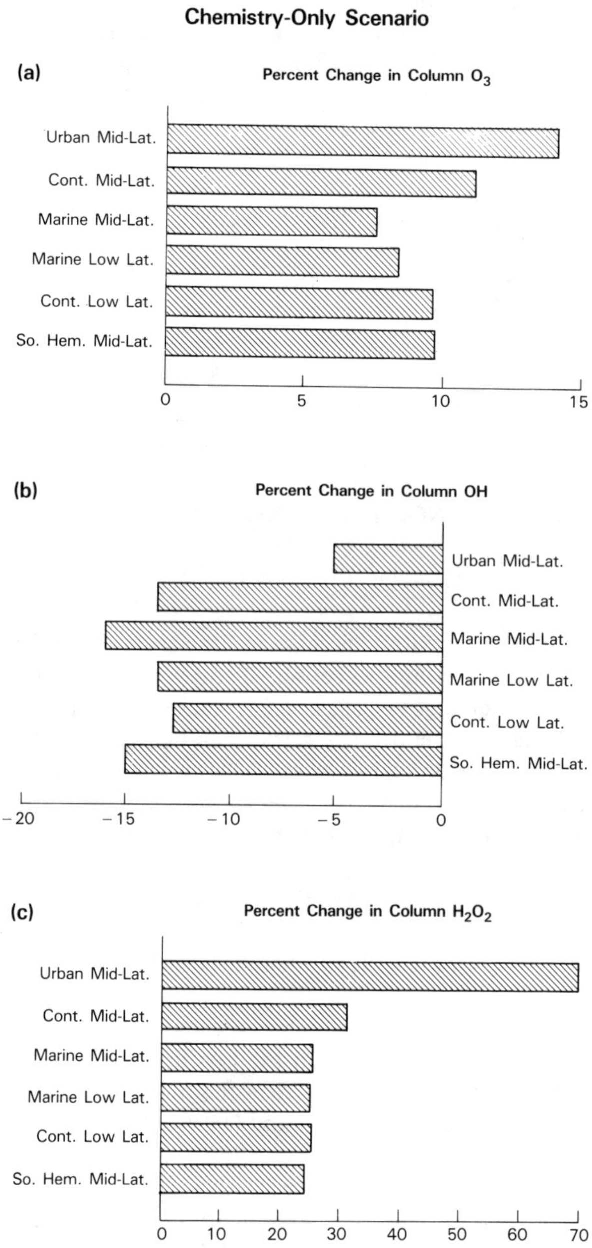

Figure 1. Fractional change in tropospheric column amount for chemistry-only scenario, $(X_{2035} - X_{1985})/X_{1985}$, for six chemically coherent regions. (a) Ozone; (b) OH; (c) H_2O_2. Scenario is described in Table 2.

Table 3. Oxidant Changes in Troposphere (0 to 15 km) for Three Scenarios of Global Change

Global Scenario	1985 → 2035 % Change, O_3	% Change, OH	% Change, H_2O_2
(1985, 0 km mix. rat.)	24.4 ppbv	$1.07x10^6/cm^3$	0.511 ppbv
1 (Chemistry only)	+9.8%	−13%	+26%
2 (Chemistry + UV)	+5.7%	−5.9%	+32%
3 (Chemistry + UV + more H_2O)	+2.0%	−1.0%	+37%

Note: mix. rat. = mixing ratios.

In the Urban case, the $[HO_2]/[OH]$ ratio increases 60% at 0 km, giving an H_2O_2 surface increase of more than 100% from 1985 to 2035.

Estimates of globally averaged O_3, OH, and H_2O_2 are made by adding the individual amounts of each oxidant (column integrated, 0 to 15 km) with the area weighting factors in Table 1. The total 1985 and 2035 budgets are compared and percent changes are summarized in Table 3.

The second scenario changes (chemistry + UV) are also summarized in Table 3. From 1985 to 2035, O_3 increases and OH decreases, but not as much as in the first scenario. This is because more UV means more OH and a loss of O_3 (Reactions 7 and 8) which partially offset the effects of CH_4 and CO forcing. An exception is in the Urban region near the surface where OH and O_3 both increase. This means that more stratospheric O_3 depletion could lead to an increase in urban smog. The chemistry and UV scenario effects on surface O_3 in 2035 appear in Figure 2. The UV effect reinforces the chemistry-only tendency to increase H_2O_2 (see Table 3).

When more water vapor is added to the chemistry and UV perturbations in the second scenario, OH and H_2O_2 are further enhanced at the expense of O_3. Globally, O_3 and OH changes are practically negligible, although H_2O_2 will increase even more, 37% globally (see Table 3).

DISCUSSION

There are practical consequences of the changes to O_3, OH, and H_2O_2 presented here. For example, the global increase estimated for the chemistry-only change in tropospheric O_3 (+ 10 to 15% from 1985 to 2035) amounts to 1 to 1.5% of the total O_3 column, assuming that tropospheric O_3 is 10% of total O_3 over most of the earth. Stratospheric ozone depletion is estimated by most scientists to be on the order of 5% from 1985 to 2035[23] and these changes will be monitored largely by looking at the total column with ground-based and

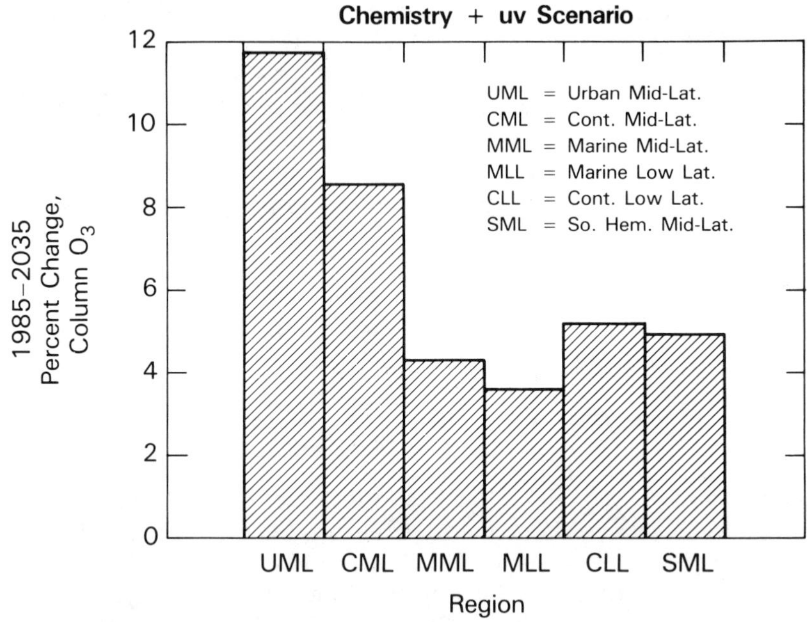

Figure 2. Fractional change in tropospheric O_2 for chemistry + UV scenario. More UV penetration into troposphere causes ozone to photodissociate: $O_3 + h\nu \rightarrow O_2 + O(^1D)$. For each region O_3 increases are less than in chemistry-only perturbation (Figure 1a).

satellite measurements.[24] The tropospheric increase is of comparable magnitude to the expected stratospheric O_3 loss, which could complicate detection of stratospheric ozone trends.

For OH, global losses in the chemistry-only change would mean slower rates of nitric and sulfuric acid formation (presumably good), but longer lifetimes for the numerous natural and anthropogenic trace gases that are oxidized by reaction with OH. These include most species with an extractable hydrogen atom (because H_2O is formed in the process) and many of the proposed CFC (chlorofluorocarbon) substitutes. The latter are rated in terms of their "ozone-depleting-potential" and this potential will increase if global OH decreases.

Higher levels of H_2O_2 in the troposphere imply greater sulfuric acid formation in clouds and rain[25,26] when acid formation is oxidant-limited. Although global average rates of H_2O_2 increase are more than 25%, our calculations show increases of >100% predicted for H_2O_2 in the urban boundary layer.

There are several sources of uncertainty in this study. First, global averages have been obtained from a series of one-dimensional calculations, which are no substitute for a fully coupled photochemical-dynamics model that simulates transport of ozone. Second, feedbacks, among tropospheric ozone-stratospheric ozone-climate changes are neglected. We also have not simulated other perturbations that may affect tropospheric chemistry if a global warming occurs, e.g.,

Table 4. Region-Specific Scenarios for Trace Gas Changes (1985 to 2035) (% Annual Change in Flux) (H = High; L = Low[a])

Scenario	1	2	3	4	5	6	7	8
CH_4[b]	L	L	L	L	H	H	H	H
	+0.58	+0.70	+0.67	+0.79	+0.99	+1.2	+1.3	+1.3
NO	L	H	L	H	L	H	L	H
Urban-1 and Continental-1	−1.1	+0.04	−0.88	+0.20	−1.1	+0.04	−0.88	+0.20
Urban-2 and Continental-2	0.0	+1.3	+0.15	+1.4	0.0	+1.3	+0.15	+1.4
CO	L	L	H	H	L	L	H	H
Urban-1 and Continental-1	−1.6	−1.3	−0.66	−0.54	−1.6	−1.3	−0.66	−0.54
Urban-2 and Continental-2	−0.21	+0.16	+0.10	+0.33	−0.21	+0.16	+0.10	+0.33

[a]Reference 31.
[b]Fractional change in flux is uniform for all regions. Intial fluxes (for 1985) of Urban-1 and Urban-2 are the same. 1985 Fluxes of Urban-1 and Continental-1 differ.

cloud cover changes, precipitation, stratosphere-to-tropospheric influxes of ozone and odd nitrogen, and frequency rate and distribution of stagnation episodes.[27] We note, however, that the rate of change we calculate for tropospheric O_3 is close to (or less than) what has been observed over the past few decades.[1,3]

Perhaps the most obvious uncertainty is in the assumed scenarios. The budgets of CH_4 and CO include multiple sources,[28-30] most of them directly or indirectly affected by human activity. It is difficult to predict future emissions given the number of factors that influence energy demand (hence, fossil fuel combustion) and land use (i.e., vegetative sources of CH_4 and CO). Some regions may institute emissions controls; others may allow emissions to increase largely unchecked. We have performed a series of model simulations for chemically coherent regions treating changes in CH_4/CO/NO emissions parametrically.[10] In those calculations, Urban and Continental regions (see Table 1) were subdivided into regions that will grow with emissions controls ("Urban-1" and "Continental-1") and ones that will restrict emissions less or not at all ("Urban-2" and "Continental-2"). Emissions projections are taken from econometric models for various geopolitical regions[31,32] and are adapted to the chemically coherent regions. A set of eight different CH_4/CO/NO emissions scenarios is run in each region (Table 4). For a given region, the range of perturbed mixing ratios over the eight scenarios is a measure of the effect of uncertain emissions. An example of NO_x and CO perturbations from 1985 to 2035 for two Urban regions and two Continental regions appears in Figure 3. The corresponding surface O_3 responses are shown in Figure 4. Note, for both regions, the wide range of changes in CO and NO depending on whether scenarios are characterized by positive or negative growth rates of each trace gas. In addition, the mean NO_x and CO increases for Urban-1

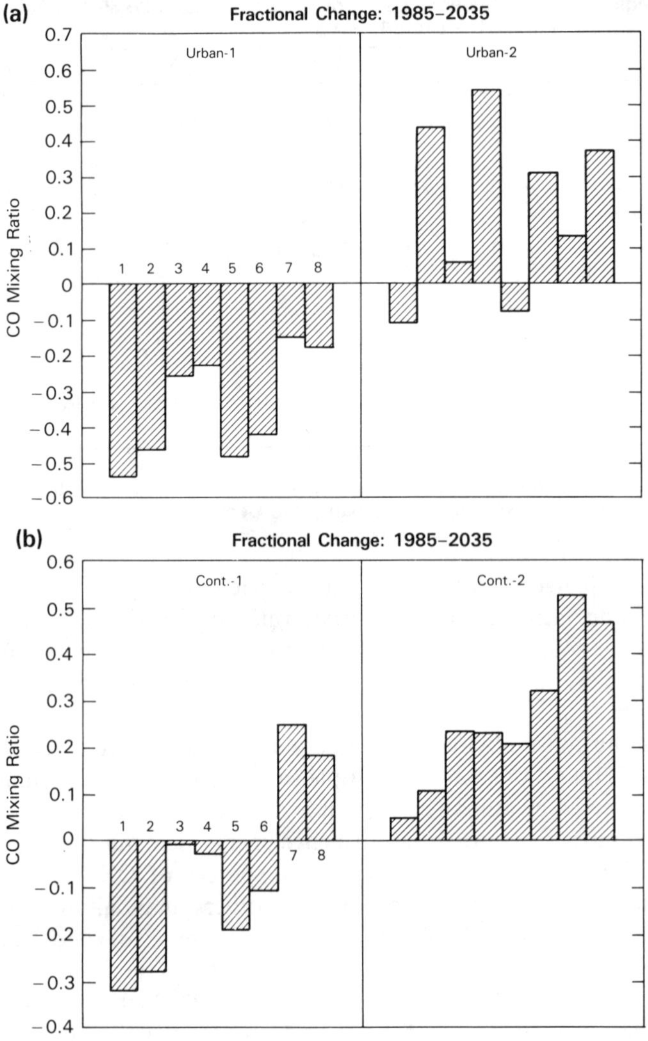

Figure 3. Fifty year fractional change in surface amounts of CO (a and b) and NO_x (c and d) for Urban and Continental regions. The scenario labels 1 to 8 are summarized in Table 4. Urban-1 and Urban-2 have same 1985 composition, but the absolute increase in CO and NO emissions differ between Urban-1 and Urban-2 within a given scenario. The same is true for Continental-1 and Continental-2.

(controlled emissions) are much less than for Urban-2 (higher CO and NO growth rates). Mean surface O_3 and OH changes differ accordingly, with Scenario 8 for Urban-2 showing a 33% O_3 increase. When surface O_3 ozone concentrations in all regions are averaged to give a global change for ozone in Scenario 8, the increase is ~25%, whereas several individual regions show O_3 increasing more than 40%.

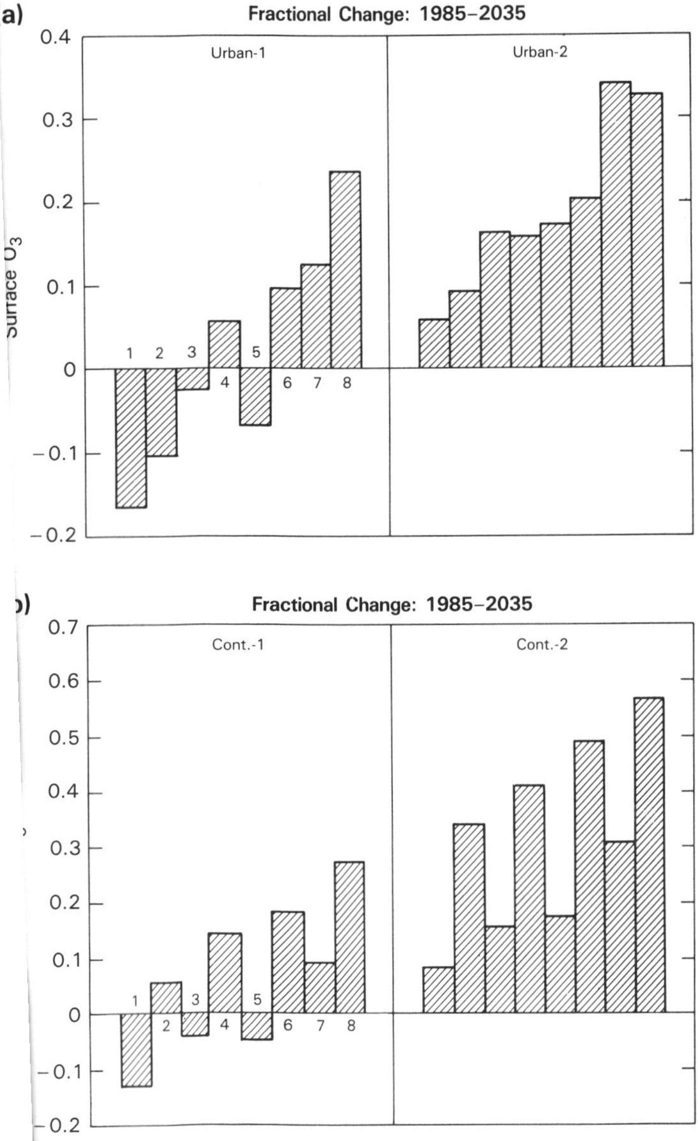

Fifty year surface O_3 changes for northern mid-latitude regions. (a) Urban; (b) Continental.

2, 4, 6, and 8 (high NO growth rates). This is because for the same rate of NO increase, the lower NO and O_3 region forms additional fficiently.[33]

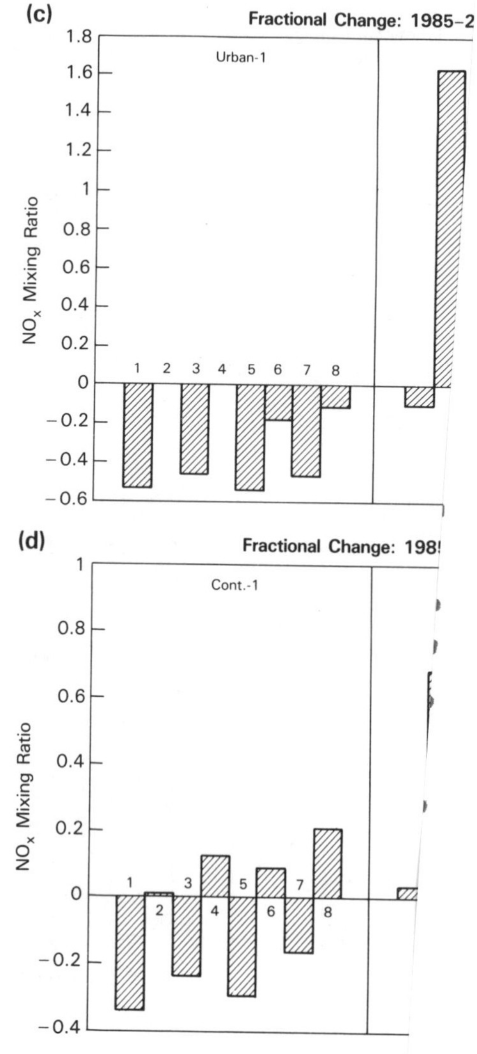

Figure 3 continued.

Figure 4 shows less than 5% O_3 increase i
narios for Urban-1. This is because Urban-1
growth rates or even decreases in emissions (
the same 1985 chemical composition as Urb
average, more than 15% by 2035. Even mor
changes in Continental-1 and Continental
affected by the same rates of CH_4/CO/NC
parts, but starting at lower O_3, NO, and
increase for these regions will be greater tha

Surface O

Figure 4.

Scenarios
fractional
O_3 more e

SUMMARY

Future changes in tropospheric ozone (and OH and H_2O_2) have been estimated with a photochemical model that simulates specific scenarios of CH_4/CO/NO, reduced stratospheric O_3 and increased water vapor, which is one possible consequence of a global warming. The period 1985 to 2035 has been simulated. The principal results are

1. Assuming chemical changes only, i.e., CH_4 and CO emissions continuing to increase at current rates of 0.5 to 1%/year, we estimate that tropospheric O_3 will increase globally ~10%, with OH decreasing ~10% and H_2O_2 increasing ~25%.
2. Adding increased UV (or UV and water vapor) to the CH_4 and CO increases will tend to counteract O_3 and OH changes of the scenario with only CH_4 and CO increases. Hydrogen peroxide, however, will increase even more.
3. A series of regionally refined scenarios based on detailed budgets of CH_4/CO/NO shows that in regions of poorly controlled growth, boundary layer O_3 may increase 30 to 50% from 1985 to 2035; O_3 increases could be less than 10% in regions that control emissions. The calculations show that the pivotal constituent is NO, with the largest O_3 increases predicted in regions of rapid NO growth and moderate present-day NO_x levels (<1 ppbv).

ACKNOWLEDGMENTS

The author thanks Mary Ann Huntley for programming assistance. This work was supported by the NASA Tropospheric Chemistry Program and the U.S. Environmental Protection Agency through Interagency Agreement #DW80933081-01-0.

REFERENCES

1. Logan, J. A. "Tropospheric Ozone: Seasonal Behavior, Trends, and Anthropogenic Influences," *J. Geophys. Res.* 90:10,463–10,482 (1985).
2. Bojkov, R. D. "Surface Ozone During the Second Half of the Nineteenth Century," *J. Clin. Appl. Meteor.* 25:343–352 (1986).
3. Volz, A., and D. Kley. "Evaluation of the Montsouris Series of Ozone Measurements Made in the Nineteenth Century," *Nature* 332:240–242 (1988).
4. Fishman, J., V. Ramanathan, P. J. Crutzen, and S.-C. Liu. "Tropospheric Ozone and Climate," *Nature* 282:818–820 (1979).
5. Thompson, A. M., and R. J. Cicerone. "Atmospheric CH_4, CO, and OH from 1860 to 1985," *Nature* 321:148–150 (1986).
6. Thompson, A. M., and R. J. Cicerone. "Possible Perturbations to Atmospheric CO, CH_4, and OH," *J. Geophys. Res.* 91:10,853–10,864 (1986).
7. Thompson, A. M. "Atmospheric Chemical and Climate Change: Possible Effects on Tropospheric Ozone," in *Ozone in the Atmosphere,* R. Bojkorand and P. Fabian, Eds., (Hampton, VA: Deepak Publishing Co., 1989), pp. 580–583.
8. Thompson, A. M., M. A. Owens, and R. W. Stewart. "Sensitivity of Tropospheric

Hydrogen Peroxide to Global Chemical and Climate Change," *Geophys. Res. Lett.* 16:53–56 (1989).
9. Thompson, A. M. "Effects of Atmospheric Chemical and Climate Change on Tropospheric Ozone," *Ozone Sci. Engineer.* 12:177–194 (1990).
10. Thompson, A. M., M. A. Huntley, and R. W. Stewart. "Perturbations to Tropospheric Oxidants, 1985–2035: Calculations of Ozone and OH in Chemically Coherent Regions," *J. Geophys. Res.* 95:9829–9844 (1990).
11. Craig, H., and C. C. Chou. "Methane: The Record in Polar Ice Cores," *Geophys. Res. Lett.* 9:1221–1224 (1982).
12. Rasmussen, R. A., and M. A. Khalil. "Increases in the Concentration of Atmospheric Methane," *Atmos. Environ.* 15:883–886 (1981).
13. Stauffer, B., G. Fischer, A. Neftel, and H. Oeschger. "Increase of Atmospheric Methane Recorded in Antarctic Ice Core," *Science* 229:1386–1388 (1985).
14. Khalil, M. A. K., and R. A. Rasmussen. "Carbon Monoxide in the Earth's Atmosphere: Increasing Trend," *Science* 224:54–56 (1984).
15. Hansen, J., I. Fung, A. Lacis, D. Rind, S. Lebedeff, R. Ruedy, G. Russell, and P. Stone. "Global Climate Changes as Forecast by Goddard Institute for Space Studies Three-Dimensional Model," *J. Geophys. Res.* 93:9341–9364 (1988).
16. Hameed, S., J. P. Pinto, and R. W. Stewart. "Sensitivity of the Predicted CO-OH-CH_4 Perturbations to Tropospheric NO_x Concentrations," *J. Geophys. Res.* 84:763–768 (1979).
17. Thompson, A. M., R. W. Stewart, M. A. Owens, and J. A. Herwehe. "Sensitivity of Tropospheric Oxidants to Global Chemical and Climate Change," *Atmos. Environ.* 23:519–532 (1989).
18. Hameed, S., and J. Dignon. "Changes in the Geographic Distributions of Global Emissions of NO_x and SO_x from Fossil-Fuel Combustion Between 1966 and 1980," *Atmos. Environ.* 22:441–449 (1988).
19. Logan, J. A. "Rural Ozone in the United States," *J. Geophys. Res.* 94:8511–8532 (1989).
20. Seiler, W., and J. Fishman. "The Distribution of Carbon Monoxide and Ozone in the Free Troposphere," *J. Geophys. Res.* 86:7255–7265 (1981).
21. Seiler, W., H. Griehl, and E. Halliday. "The Seasonality of CO Abundance in the Southern Hemisphere," *Tellus* 36B:219–231 (1984).
22. Lenschow, D. H., R. Pearson, Jr., and B. B. Stankov. "Measurements of Ozone Vertical Flux to Ocean and Forest," *J. Geophys. Res.* 87:8833–8838 (1982).
23. World Meterological Organization. "Atmospheric Ozone: Assessment of Our Understanding of the Processes Controlling its Present Distribution and Change," WMO Report 16, Geneva (1986).
24. NOAA FCM. *National Plan for Stratospheric Monitoring: 1988–1997*, NOAA Publication FCM-P17-1989, Washington, D.C. (1989).
25. Penkett, S. A., B. M. R. Jones, K. A. Brice, and E. J. Eggleton. "The Importance of Atmospheric Ozone and Hydrogen Peroxide in Oxidizing Sulphur Dioxide in Cloud and Rain Water," *Atmos. Environ.* 13:123–137 (1979).
26. Chameides, W. L., and D. D. Davis. "The Free Radical Chemistry of Cloud Droplets and Its Impact Upon the Composition of Rain," *J. Geophys. Res.* 87:4863–4877 (1982).
27. Rind, D., R. Suozzo, N. K. Balachandran, and M. J. Prather, "Climate Change in the Middle Atmosphere: The Doubled CO_2 Climate," *J. Atmos. Sci.* 47:475–494 (1990).

28. Ehhalt, D. H. "How Has the Atmospheric Concentration of Methane Changed?," in *The Changing Atmosphere*, F. S. Rowland and I. S. A. Isaksen, Eds. (New York: John Wiley & Sons, Inc., 1988), pp. 25–32.
29. Crutzen, P. J., I. Aselmann, and W. Seiler. "Methane Production by Domestic Animals, Wild Ruminants, Other Herbiforous Fauna, and Humans," *Tellus* 38B:271–284 (1986).
30. Logan, J. A., M. J. Prather, S. C. Wofsy, and M. B. McElroy. "Tropospheric Chemistry: A Global Perspective," *J. Geophys. Res.* 86:7210–7254 (1981).
31. ICF Incorporated. "Global Macro Energy Model: Summary Paper," available from U.S. Environmental Protection Agency, Washington, D.C. 20460.
32. EPA. "Policy Options for Stabilizing Global Climate," Draft Report to Congress, U.S. Environmental Protection Agency, Washington, D.C. 20460, 1989.
33. Liu, S. C., M. Trainer, F. C. Fehsenfeld, D. D. Parrish, E. J. Williams, G. Hubler, and P. C. Murphy. "Ozone Production in the Rural Troposphere and the Implications for Regional and Global Ozone Distributions," *J. Geophys. Res.* 92:4194–4207 (1987).

CHAPTER 5

A Three-Dimensional Photochemical-Aerosol Model for Episodic and Long-term Simulation: Formulation and Initial Application in the Los Angeles Basin

T. W. Tesche and Dennis McNally

INTRODUCTION

Traditionally, urban- and regional-scale Eulerian photochemical models have been applied to periods of one to a few days in length. Photochemical simulations have been limited to a few prototype days as a result of (1) the 1 hr averaging time of the federal ozone standard and (2) the computer requirements for direct calculation of seasonal or annual averages. While this episodic focus in grid-based modeling is understandable, some emerging problems now require longer duration simulations.

In ozone State Implementation Planning (SIP), for example, there is no *a priori* guarantee that the emission control requirements developed for the "worst case" meteorological scenario will be the most stringent. Modeling studies by the EPA have shown that the emission control requirements for ozone attainment in St. Louis may be greater for less severe conditions, compared with the worst case design episode.[1,2] Decision makers need assurance that the ozone precursor control requirements developed on the basis of one or two modeling episodes will provide for attainment under all weather conditions occurring during the annual cycle. Through simulation of all meteorological conditions occurring throughout the year, it is possible to identify directly those periods that impose the most stringent ozone control requirements.

The formation of secondary sulfate, nitrate, ammonium, and organic aerosols (all components of atmospheric fine particulate below 10 μm, i.e., PM-10) is influenced by many of the same physicochemical processes governing ozone. Since the federal PM-10 standard includes an annual average, aerosol formation processes need to be treated over all pertinent time scales within the annual period. Empirical models have been developed to estimate annual average PM-10 control requirements based on linearized chemistry and meteorology,[3] but the important nonlinear features of photochemical oxidant and aerosol formation are lost in these simple approaches.[4] Direct numerical simulation of the photochemical oxidant-secondary aerosol processes is the most rigorous approach to dealing with the annual PM-10 estimation problem.

Finally, toxic contaminants such as benzene, emitted from a variety of urban sources, need to be modeled on an annual-average basis in order to support quantitative cumulative health risk assessments. Conventional Gaussian dispersion models are limited in their ability to simulate accurately and precisely source-receptor relationships during periods of nonuniform, temporally varying meteorology, particularly when chemical transformations and removal processes are important.

Limited work has been performed to extend Eulerian models to annual-average time frames. No direct methods, such as we develop in this chapter, have been attempted; however, three indirect methods have been proposed. Lamb[5] constructs a method involving numerical tracer simulations to typify meteorological transport scenarios. The scenarios are combined in a weighted manner to develop an annual average. As part of the Regional Acid Deposition Model (RADM) Program, an aggregation scheme is proposed based on statistical analysis of climatological regimes.[6] RADM modeling results from prototypical regimes are then aggregated to form the annual average. Researchers at Brookhaven National Laboratory have proposed a simple linear transport and removal model to develop a synthetic data base that then may be used to support annual-average estimation.[7] This method invokes cluster analysis.

The idea of modeling gas-phase photochemistry and secondary aerosol formation in an urban area by direct numerical simulation on an annual basis has been discussed for a number of years. In this chapter we describe current progress in the development of a Eulerian photochemical-aerosol model. The initial model framework has been developed and some preliminary results from a 30-day ozone simulation in the Los Angeles air basin are presented. Research now is aimed at installing the aerosol formation component of the model.

Our initial 1-month ozone simulation results are encouraging. The model's gross error in simulating hourly averaged ozone concentrations is approximately 60%, which may be compared with the 35 to 40% errors commonly found with episodic ozone simulations. Good qualitative agreement is achieved between the Annual Average Urban Airshed Model and the Urban Airshed Model (UAM) for the June 5 to 7, 1985 episode days that were used as the basis for the South Coast Air Basin's Air Quality Management Plan (AQMP). These preliminary simulations suggest that annual-average photochemical modeling is technically and computationally feasible, providing impetus for continued model development and testing.

FORMULATION OF THE 3AM MODEL

The Annual Average Urban Airshed Model (3AM) is conceived as a three-dimensional, time-dependent, Eulerian, transport, transformation, and deposition model designed to simulate urban-scale air pollution problems over arbitrary time periods from 1 hr to 1 year. The model is based on the "semi-empirical atmospheric diffusion equation"[8] that describes mass conservation of chemically reactive and inert species within the planetary boundary layer. Modular in design, 3AM is intended for use over urban- to mesoscale domains. Once fully developed, it may be applied for either inert or linearly reactive gaseous compounds or for full photochemical oxidant reaction systems that may include secondary particulate aerosol formation. Figure 1 is a sketch of the 3AM model concept.

The 3AM model integrates a coupled set on nonlinear, partial differential species conservation equations given by:

$$\underbrace{\frac{\partial c_i}{\partial t}}_{\text{Time dependence}} + \underbrace{\frac{\partial(uc_i)}{\partial x} + \frac{\partial(vc_i)}{\partial y} + \frac{\partial(wc_i)}{\partial z}}_{\text{Advection}}$$

$$= \underbrace{\frac{\partial}{\partial x} K_H \frac{\partial c_I}{\partial x} + \frac{\partial}{\partial y} k_H \frac{\partial c_i}{\partial y} + \frac{\partial}{\partial z} K_V \frac{\partial c_i}{\partial z}}_{\text{Turbulent diffusion}} + \underbrace{R_i}_{\text{Chemical reaction}} + \underbrace{S_i}_{\text{Emissions}} + \underbrace{L_i}_{\text{Removal}}$$

where subscript i is the concentration of species i; u, v, and w are components of the three-dimensional velocity vector at each grid point in the computational domain; K_h and K_v are horizontal and vertical components of the eddy diffusivity tensor describing turbulent mass exchange processes; R_i is the rate of change of species i via chemical reactions; S_i is the emission rate of precursor species; and L_i describes the removal rate of species i due to wet and dry deposition. Included in the term R_i is the evolution of the mass concentration of aerosol species undergoing nucleation, condensation, and coagulation. The model equations are integrated forward in time using finite difference techniques.

Numerical solution of Equation 1 requires specification of initial and boundary conditions, together with time- and space-resolved descriptions of the meterology and emissions at each relevant grid point. Chemical transformations are treated with a modular chemical reaction mechanism, flexibly designed to allow simulation of inert species, gas-phase photochemical systems, and secondary aerosol formation. Several of the conceptual and computational features of the 3AM model have been adapted directly from the Urban Airshed Model (UAM). The UAM, the EPA's currently recommended guideline model for urban-scale refined ozone analyses, has been under continuous

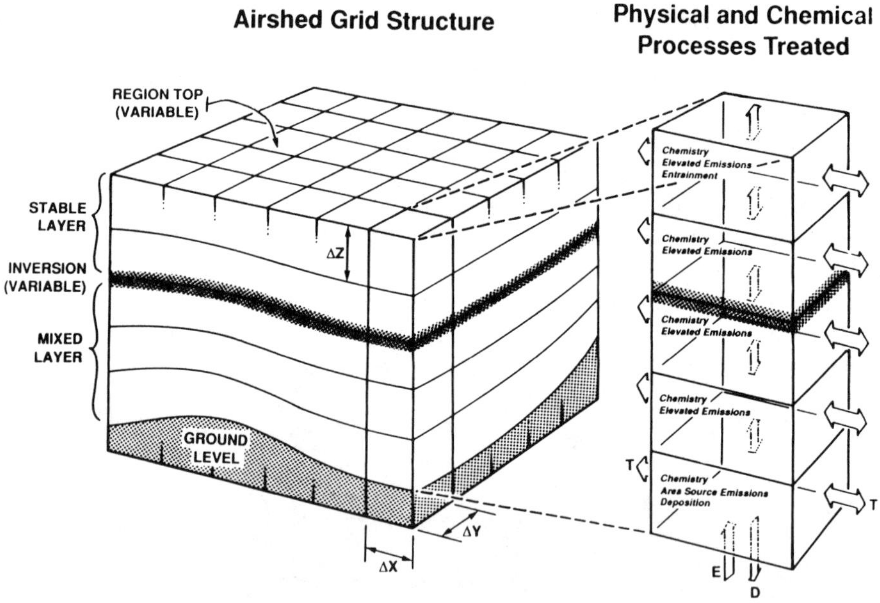

Figure 1. Sketch of the Annual Average Airshed Model (3AM) concept. $\Delta X \approx 4000$ m; $\Delta Y \approx 4000$ m; $\Delta Z \approx 250$ m; D, dry deposition; E, area source emissions; T, transport.

development and evaluation since the early 1970s.[9] A detailed description of the UAM formulation and the procedures used in its application is given elsewhere.[10] An overview of the 3AM model formulation and solution procedures is presented below.

Transport

Relevant Phenomena

The concentration dynamics of ozone, secondary aerosols, and precursor species are influenced significantly by transport processes occurring within the atmospheric boundary layer. Principal among these are the three-dimensional wind field, the intensity and spatial variation of temperature inversions, pollutant-removal processes, and the changes in atmospheric stability over the diurnal cycle. A novel feature of the 3AM is the approach for constructing transport fields internal to the simulation model, in parallel with the emissions, dispersion, and chemical transformation calculations.

Wind Fields

Three distinct conceptual approaches were considered for constructing hourly averaged, three-dimensional mesoscale wind fields: objective analysis (e.g., interpolation), diagnostic modeling, and prognostic modeling. Prognostic (or primitive equation) modeling is theoretically preferable, being based on a full description of mass, momentum, and energy conservation relationships. This method is too computationally intensive for annual average modeling, however. The approach we have adopted in 3AM is a hybridization of interpolative and diagnostic techniques. The Mountain Wind Model[1,12] employs a two-step procedure, whereby diagnostic algorithms generate an initial three-dimensional flow field that is strongly influenced by topography and boundary layer forcing. The three-dimensional, steady-state mass conservation equation is solved numerically, subject to boundary layer parameterizations of topographic lifting, channeling, convergence due to friction and thermal effects, mountain-valley wind circulations, and flow separation over ridges. The field is then objectively filtered using the Goodin et al.[13,14] interpolation routines to remove unrealistic or spurious accelerations, to ensure mass conservation, and to incorporate observational data. The diagnostic algorithms are reported elsewhere in detail.[15]

Wind fields for 3AM are calculated at the beginning of each simulation hour, based on wind speed and direction data from surface and upper air monitoring locations. The model input structure has been designed to read processed National Weather Service (NWS) data files or routine aerometric data files maintained by local reporting agencies (e.g., the South Coast Air Quality Management District).

Temperature Fields

Estimation of the gridded, hourly surface temperature field is often complicated because many regions of interest include water or mountainous terrain. This is particularly true in Los Angeles. Between these two distinct areas are the most uniform coastal plains and inland valleys, where measurement sites are more frequent. Owing to the unique physiographic characteristics of the South Coast Air Basin, we selected the surface temperature estimation technique originally proposed by Pielke and Mehring[16] and adapted to the region by Tesche and co-workers.[17] Hourly nonlinear regression equations are developed from several local meterological monitoring stations in the basin and adjacent mountains. These are then used to construct terrain-temperature regression equations. Gridded surface temperature fields are then developed from these equations using the terrain elevation data. Independent testing of this procedure using detailed meterological data collected in the San Bernardino and Santa Ynez Mountains has shown that it provides an excellent representation of the local surface temperature in regions where data are lacking or in short supply.[18]

Mixing Heights

Similar to the UAM, the 3AM uses the mixing height concept to divide the modeled domain into two regions that may differ significantly in their turbulent structures. For convective and neutral boundary layers, this division into two regions is conceptually appealing. For the nocturnal stable boundary layer, however, greater grid resolution is needed. For the present, a vertical grid mesh of four levels, with two cells below the nominal daytime mixing height and two cells above, is used. For daytime convective conditions, a modified version of the Steyn-Oke mixing height model[19] is used. We extended the original model to incorporate local features of the Southern California region and to estimate the turbulent sensible heat flux from routine NWS meteorological measurements. Nighttime mixing heights under stable conditions are calculated estimated from the algorithms of van Ulden and Holtslag.[20] Recent mixing height model intercomparison studies in Southern California[21] show these methods to be the best currently in use.

Precipitation Fields

Precipitation processes are approximated in 3AM through the use of a storm-averaged wet-removal parameter, A_i (x,y,z), that is estimated from the type and intensity of precipitation reported at NWS sites. The precipitation field parameter may be determined from reported precipitation type and intensity and as a function of altitude.[22] This empirical approach has obvious shortcomings; however, for regions such as Southern California, where precipitation is rare during the summer-autumn oxidant season, the approximation is adequate.

Meterological Scalars

The 3AM requires a number of scalar meteorological quantities to calculate atmospheric stabilities, turbulent diffusion rates, surface deposition rates, plume rise, and so on. Routine meterological observations are used together with current boundary layer parameterization schemes[20,23,24] to calculate the surface turbulent heat flux, friction velocity, Monin-Obukhov length, turbulence intensity, convective velocity scale, and the temperature scale for turbulent heat transfer. These parameters are updated hourly based on gridded surface roughness and vegetation factors and estimates of local surface wind speed and direction, early morning temperature and soundings, surface albedo, and soil moisture content.

Emissions

The 3AM requires as input a "high-resolution" inventory of emissions from all relevant anthropogenic and natural sources in the modeling region. The inventory must be spatially resolved (e.g., on a 4- to 5-km grid mesh), temporally allocated (hourly emissions estimates), and disaggregated according to nitrogen oxides (NO_x) and reactive organic gases (RO_G) chemical species. Procedures for developing photochemical modeling source inventories at this level of detail are prescribed by the EPA.[25,26]

The inventory contains the emission rates of various pollutant species for all sources in the region. It is constructed from baseline estimates of emissions of total organic gases (TOG), NO_x, particulate matter (PM), sulfur dioxide (SO_2), and carbon monoxide (CO) from point and area sources. Steps in constructing the high-resolution inventory include:

1. Source identification
2. Source and process characterization
3. Source activity level determination
4. Emission factor estimation
5. Emission rate calculation
6. Emission gridding, temporal splitting, and species allocation
7. Inventory quality assurance

Emissions are distributed temporally and spatially, including in the vertical plane, consistent with the modeling grid mesh. Point sources are located through the use of specific coordinates identifying their position in the grid region. The model uses stack data to compute plume rise for all point source emissions, based on the current hourly meterological conditions. Area source emissions are allocated to grid cells using surrogate information, such as population or industrial employment distributions, for each source category. Operating schedules for point and area sources are used for resolving temporally emissions data to specific time periods. Day-specific emissions information may also be gathered for particularly large sources so that the final inventory is more representative of conditions during specific days. Weekday-weekend, holiday, and seasonal differences in the emissions inventory are treated in an empirical manner by applying adjustment factors to the nominal summer weekday emission inventory normally developed for photochemical modeling. The appropriateness of these factors needs to be determined, since there is very little reliable information presently available in this area.

Chemistry

Given the multiple uses of the 3AM, alternative treatments of atmospheric chemistry are needed depending on whether the system of interest (1) is inert or linearly reactive (e.g., for health risk assessments); (2) involves gas-phase photochemical oxidants (principally ozone); and (3) includes the gas-phase and secondary aerosol formation processes associated with PM-10 air pollution. In its current state of development, 3AM treats inert and linearly reactive chemistry, gas-phase ozone chemistry, and highly parameterized aerosol formation. Implementation of a detailed secondary aerosol mechanism into the model is in progress. The following sections describe the existing gas-phase photochemical mechanism and the general approach in implementing secondary aerosol chemistry.

Gas-Phase Photochemistry

The Carbon Bond II (CBM-II) chemical mechanism[27] is currently used in 3AM to solve the gas-phase photochemistry portion of the reaction term R_i in Equation 1. The CBM-II chemical reaction mechanism contains 65 chemical reactions involving 27 different species. The Carbon Bond II mechanism has evolved over many years and has been subjected to extensive evaluation efforts with environmental chamber data, as well as with atmospheric data.[28] In addition to simulating the gas-phase hydrocarbon-NO_x-O_3 system, CBM-II is also capable of treating the oxidation of SO_2 and the formation of secondary aerosols in an approximate sense. The mechanism can estimate the total mass concentration of aerosol products via ten additional reactions for sulfate, nitrate, and organic aerosol. Recently, the mechanism has been refined in its treatment of sulfate and nitrate[29] through the addition of reactions treating SO_2 oxidation by OH, H_2O_2 photolysis, H_2O_2 reaction with OH, heterogeneous hydrolysis of N_2O_5, and CH_4 reaction with OH.

Although the CBM-II mechanism is well tested and computationally efficient, it has two main drawbacks. First, it does not represent the current state-of-the-science of atmospheric chemistry embodied in mechanisms such as Statewide Air Pollution Research Center (SAPRC)[30] and Carbon Bond IV (CBM-IV).[31] So far, there have been no published intercomparisons between CBM-II and CBM-IV, for example, with atmospheric data; thus, the degree of error introduced with the older mechanism is uncertain. (These tests with the June 5 to 7, 1985 data base are underway for the Los Angeles basin.) A second drawback, common to both versions of the CBM chemistry, is the approach taken in coding the chemistry solver. The Carbon Bond Mechanism implementation (CMB-II and CBM IV) makes use of the steady-state approximation for fast-reacting radical species (e.g., OH, HO_2, O). The solution scheme for the resulting nonlinear algebraic equations is "hard-wired" into the code, precluding rapid inclusion of additional species and reaction sets in the mechanism. Both of the above limitations are being eliminated in the 3AM development by switching to the CBM-IV mechanism and by implementing a much more general chemistry solution method.[32] Because the computational burden with CBM-IV is expected to quadruple, conversion to the more flexible SAPRC is also being considered.

Secondary Aerosol Chemistry

In the Los Angeles basin, the approximate composition of aerosol is 20% sulfates, 25% nitrates, 10% elemental carbon, 20% organic carbon (primary and secondary), and 25% soil, metals, and water.[32] The gas-phase portion of the 3AM simulates the temporal and spatial distributions of primary and secondary gaseous species, including the production rates of nitric and sulfuric acid. Concentrations of sulfate, nitrate, ammonium, and organic aerosols may be calculated based, in part, upon the gas-phase predictions. Other necessary descriptions are the appropriate thermodynamic representations, emission rates of ammonia, and the temperature and relative humidity of the atmosphere.

Rigorous modeling of multicomponent aerosols requires treatment of several processes: new particle formation by nucleation, growth or evaporation due to gas-to-particle or particle-to-gas conversion and coagulation, aerosol injection, and removal. A detailed mathematical framework for describing these dynamic phenomena is provided by the so-called "General Dynamic Equation" (GDE),[8] and several numerical modeling approaches have been developed from this starting point.[29,33,35,36]

We have adopted a simpler "engineering approach" for implementation of the aerosol chemistry.[29,37] The aerosol size distribution method uses a "sectional approach", originally developed to simulate aerosol dynamics in power plant plumes. The aerosol size distribution is represented by several step functions and the resulting sectional GDE is solved approximately. The sectional model, coupled with an aerosol chemistry module, permits the chemical composition of the aerosol to be estimated as a function of its size.

In addition to solving the sectional representation of the GDE, a thermodynamic calculation is used to partition volatile compounds such as NH_3 and HNO_3 between gaseous and aerosol phases. Three contemporary models are available: KEQUIL,[38] EQUIL,[39] and MARS.[40] From comparative evaluations, the MARS model has proven to be an accurate and computationally efficient thermodynamic aerosol model for predicting the quantity and composition of secondary species containing sulfate, nitrate, and ammonium compounds. It may also be used to simulate the aqueous-phase oxidation of SO_2 to sulfate by H_2O_2, O_3, and O_2. MARS predicts the chemical composition of multiphase aerosols containing water, $(NH_4)_2 SO_4$, $NH_4 HSO_4$, $(NH_4)_3 H(SO_4)_2$, $H_2 SO_4$, HNO_3, and $NH_4 NO_3$. Treatment of organic aerosol formation will be explored following the method of Grosjean and Seinfeld.[41]

Inert and Linearly Reactive Species

3AM can simulate up to 13 inert or linearly reactive species, as might be required for air quality permitting or health risk assessment studies. The transport, diffusion, and removal processes are treated in the same manner as the gas-phase and aerosol species. Detailed performance evaluations of the basic transport and dispersion algorithms used in 3AM have been reported elsewhere,[42] using more than two dozen multiple tracer experiments.

Deposition

Dry deposition removal is incorporated into the 3AM using the deposition velocity concept. The uptake flux, F_{di}, of pollutant i at the surface, is proportional to the ground-level concentration, C_{gi}. The proportionality constant is the effective deposition velocity, V_{di}. Mathematically, the flux can be written as follows:

$$F_{di} = V_{di} \times C_{gi} \tag{2}$$

Pollutant uptake takes place in three sequential stages:

- Transport to just above the surface by advection and turbulent diffusion
- Transport to the surface itself by processes that are influenced by the shape of the surface
- Absorption, adsorption, or chemical reaction at the surface

The deposition velocity is calculated from:

$$V_{di} = \frac{1}{1/(r_a + r_b + r_c)} \tag{3}$$

where r_a is the aerodynamic resistance (common to all gases) between a specified height and the surface, r_b is the quasi-laminar sublayer resistance (dependent upon molecular diffusivity), and r_c is the bulk surface resistance. The aerodynamic (i.e., transport) and viscous sublayer resistances are calculated by conventional empirical relationships.[43-45] More recent methods summarized by Walcek and co-workers[46] are being considered.

The treatment of surface resistance was developed by Killus et al.,[47] who estimated composite surface uptake velocities for various urban land use types and reference uptake velocities for each pollutant species. This approach is similar to the more refined method in the Regional Acid Deposition Model (RADM).[48] The recent parameterizations of Wesely[49] are being examined to assess their suitability for replacing the existing lookup table methodology for estimating deposition velocities for different species as a function of season and land use type.

Wet removal processes may be parameterized according to the following relationship:

$$W_i\,(x,y,z) = A_i\,(x,y,z) \times C_i\,(x,y,z) \tag{4}$$

where A_i is an empirical, local scavenging coefficient, dependent on pollutant species i and precipitation intensity.[50] The storm scavenging parameter value depends on storm type and precipitation amounts, information that can be inferred from NWS observations. The removal flux, W_i, is incorporated directly into the governing species conservation equations of the model, instead of being treated as a surface boundary condition (e.g., the dry deposition rate.) Procedures developed by Scott[22] are being considered for estimating the scavenging coefficients as a function of storm characteristics.

Numerical Methods

The multidimensional, nonlinear, and coupled character of the photochemical model equations requires that the full problem be divided into constituent parts using the splitting algorithm.[51] A four-step numerical solution procedure results. The first two steps solve the horizontal advection-diffusion problem; the third step treats vertical transport (advection and diffusion) and emissions. The fourth step includes the contributions of chemical reactions.

There have been many comparative studies of horizontal advection schemes in the last decade.[52–58] In the present version of the 3AM, the Smolarkiewicz[57] advection scheme is used. It is positive-definite, forward in time, and has been shown to be superior in accuracy and economy to the SHASTA scheme previously used in the UAM. The vertical transport and emissions step (Step 3) is solved with an implicit, finite difference technique to eliminate stability constraints on the size of the time-step that might otherwise arise if the vertical grid spacing becomes relatively small at some time during the simulation. The chemistry integration step (Step 4) used for the Carbon Bond II (CBM-II) mechanism employs a modified Crank-Nicholson difference scheme, yielding a set of nonlinear algebraic equations. A Newton iterative procedure is used to solve the resulting equation set.

Implementation of the Carbon Bond IV (CBM-IV) chemistry (currently being tested on the June 5 to 7, 1985 South Coast Air Basin (SOCAB) data base) uses a different chemistry solver. It is a modified predictor-corrector scheme[32,59] and is presently installed in the EPA's Regional Oxidant Model (ROM) and the Photochemical Box Model (PBM).

Computational Features

Initial development and testing of 3AM was performed on a SUN-4/280 workstation. A 30-day photochemical simulation on the SUN-4 required about one CPU week. Simulations of individual toxic air compounds (particularly if they are unreactive) are expected to require significantly less CPU time. Under support from the National Science Foundation, the model development work has been moved to the CRAY XM-P 48 supercomputer at the National Center for Supercomputer Applications (NCSA), at the University of Illinois, where vectorization of the model is underway.

PRELIMINARY APPLICATION TO THE LOS ANGELES BASIN

The South Coast Air Basin was selected as the prototype region for model development and initial testing because of the availability of meteorological, emissions, air quality, and demographic data bases directly suited to grid-based modeling studies. The initial simulations were performed for the period June 1 to 30, 1985 because this encompasses the 3-day episode (June 5 to 7) used by the South Coast Air Quality Management District (SCAQMD) as the basis for their updated Air Quality Management Plan. The 3AM modeling domain consists of a 34 × 20 grid mesh with 5-km grid resolution (shown in Figure 2). In comparison, the SCAQMD UAM modeling was carried out on a 64 × 30 grid, also with 5-km resolution. Our grid fits within and is aligned with the SCAQMD grid.

Scope of Simulation

The first test of 3AM was a simulation of the month of June 1985 in a "hands-off" mode. The model was initialized at 0000 PST on June 1 and was allowed to run uninterrupted. This is quite unlike UAM modeling exercises in which model input files are frequently modified, based on diagnostic analysis of the output predictions, in order to improve ozone predictions at particular locations and during specific time periods within the simulation. Since the user cannot "intervene" in a 3AM calculation (because the hourly meterological files are computed in-step with the overall simulation), the accuracy and precision of the model results will be somewhat less than that achievable with the UAM in episodic applications. However, this hands-off operation eliminates potential arbitrariness and controversy associated with input file modifications made to improve ozone predictions.

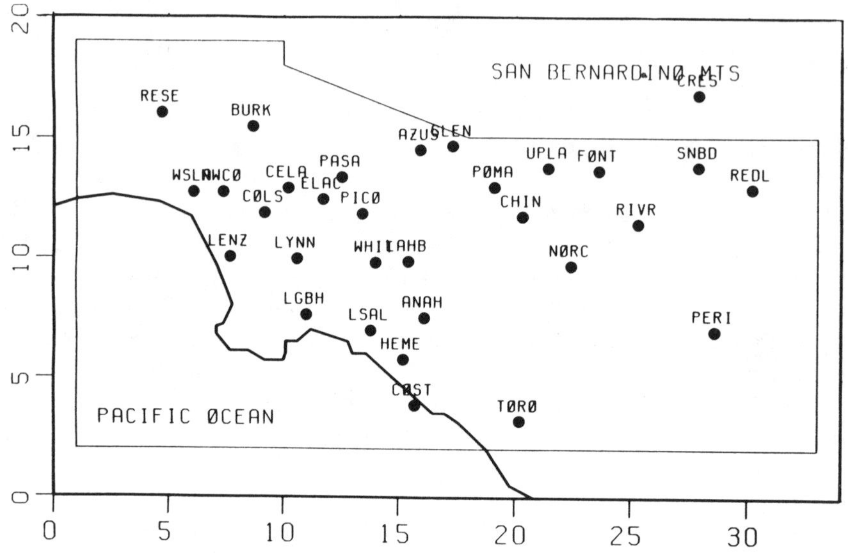

Figure 2. Modeling region of the 3AM application to the South Coast Air Basin. (Names of the air monitoring stations are given by abbreviations. Grid region consists of a 20 × 34 array of 5-km cells.)

Figure 3 presents hourly time series of predicted and observed ozone concentrations at the Pasadena monitoring station for the first 3 weeks of the simulation. Pasadena is a mid-basin site and commonly experiences high oxidant and PM-10 concentrations during air pollution episodes in the basin. The model replicates the diurnal rise and fall of ozone levels throughout the period. There is a tendency to overestimate ozone in the early morning and late evening hours at Pasadena, while midday, the ozone maxima are typically underestimated. During the period of June 5 to 8 (encompassing the SCAQMD design episode), the model reproduces the ozone peaks fairly well. The maximum value (32 parts per hundred million, pphm) is underestimated by 23% on June 6.

Figures 4, 5, and 6 present ground level ozone isopleths on June 5 to 7 between the hours of 1500 and 1600 PST, respectively. For each day, two

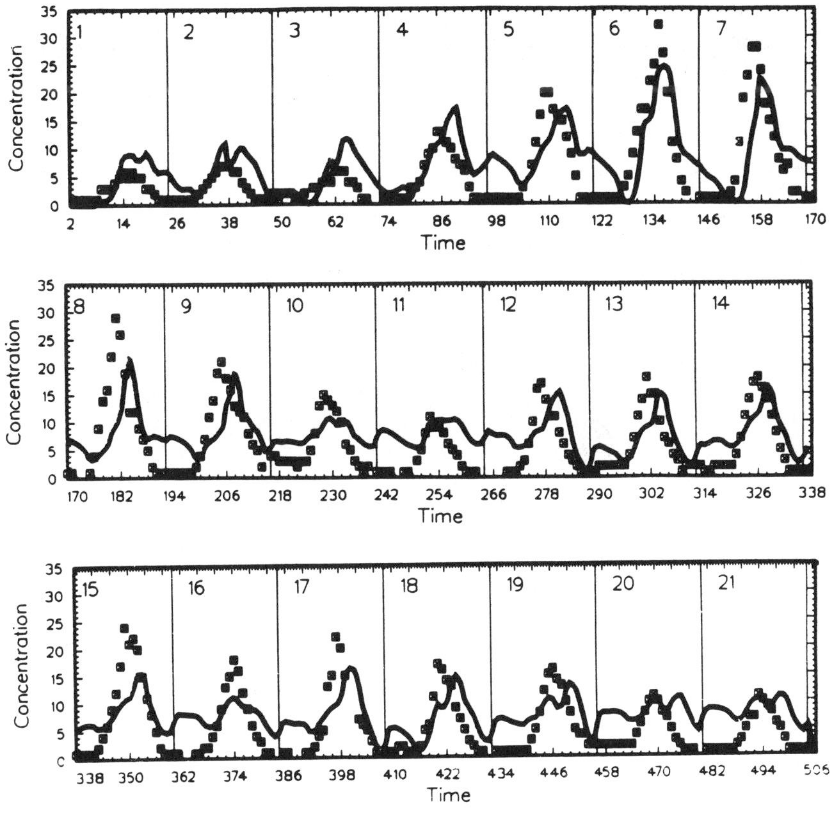

Figure 3. Predicted and observed ozone concentration time series for the first 3 weeks of June 1985 at the Pasadena Monitoring Station. (The ozone concentrations are in units of pphm. Observed hourly concentrations are identified by the boxes and 3AM model predictions are fitted with a smooth curve.)

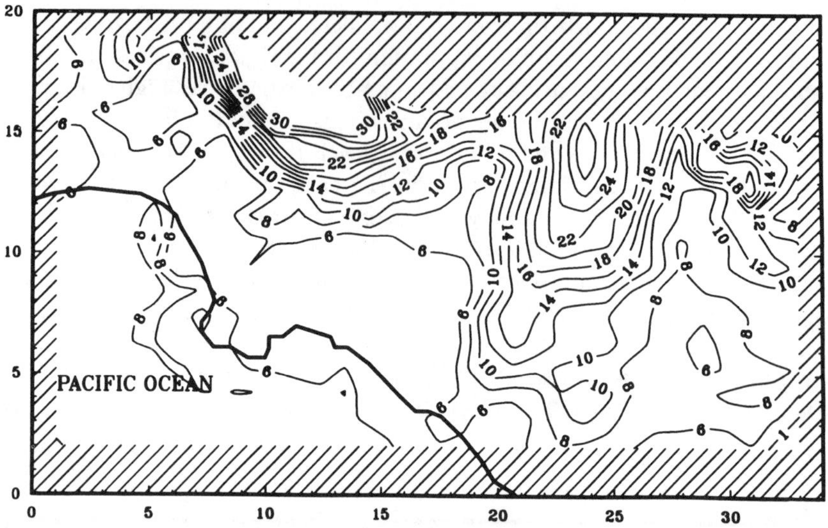

Figure 4. Predicted ground-level isopleth on June 5, 1985 between the hours of 1500 and 1600 PST. (Ozone concentrations are plotted in pphm.)

regions of high ozone are simulated, one in the San Fernando Valley and the other in the mid- to eastern basin. Concentrations equal to or greater than 30 pphm are predicted in the northeastern portion of the San Fernando Valley in the vicinity of the San Gabriel Mountains. To the east, a southwest-

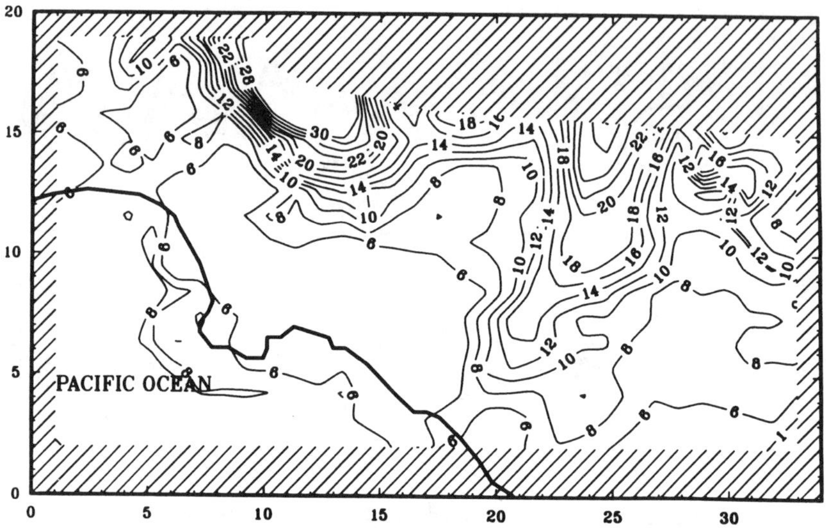

Figure 5. Predicted ground-level Ozone isopleth on June 6, 1985 between the hours of 1500 and 1600 PST. (Ozone concentrations are plotted in pphm.)

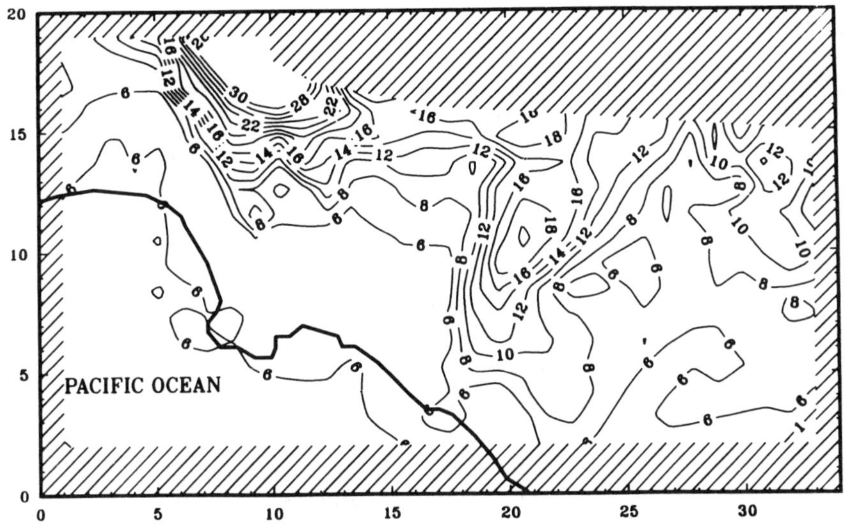

Figure 6. Predicted ground-level ozone isopleth on June 7, 1985 between the hours of 1500 and 1600 PST. (Ozone concentrations are plotted in pphm.)

northeast oriented ozone cloud is simulated over the Chino-Upland-Riverside region. Here, the ozone peaks reach 26 to 28 pphm or more.

We do not infer too much from the simulation results obtained to date because they are preliminary, and more testing and refinement of the 3AM preprocessor algorithms is necessary. Also, a reliable procedure for scaling weekday emissions to weekend and holiday periods is needed. Nevertheless, these results are encouraging. The performance of the 3AM model in simulating hourly ozone concentrations compares favorably with typical performance evaluation results obtained with the UAM. In Tesche's[11] review of over 80 photochemical grid model evaluation studies, the overall accuracy of hourly averaged ozone predictions, paired in time and space, is 35 to 40%. The episodic models (e.g., UAM) are biased low by about 10%. In the first 3AM test, the overall error is approximately 60% and bias is –30%. The 3AM results reveal many of the important features of the regional ozone distributions simulated by the UAM in the SCAQMD's June 5 to 7 episodic modeling analysis. Both models predict high concentration ozone "clouds" in about the same geographic subareas, and the magnitudes of the ozone concentrations are comparable. For the 3-day episode, the UAM gross errors varied between 33 to 38% and the bias ranged between –12 to –29%.

The preliminary results show no evidence of unusual pollutant accumulation in the modeling region. The model predicts high ozone on the days when elevated concentrations were reported, and on more moderate days when pollution levels are lower, predicted concentrations are correspondingly lower.

SUMMARY

The formulation of a new three-dimensional, Eulerian photochemical-aerosol model, currently under development, has been described. A preliminary version of the 3AM, which treats inert pollutants and gas-phase ozone photochemistry, has been applied to the South Coast Air Basin. Using only routinely available air quality and meterological data files from the South Coast District and the NWS, the model was exercised for the month of June 1985, to calculate hourly ozone concentrations throughout the basin.

Comparing hourly ozone predictions and observations (paired in time and space) at nearly 40 monitoring stations, the model's overall bias (–30%) and gross error (60%) are not dissimilar from (1) historical UAM performance statistics (–10% and 35 to 40%, respectively) and (2) the June 5 to 7, 1985 UAM results (–12 to –29 and 33 to 38%, respectively) obtained in the SCAQMD AQMP modeling analysis. The 3AM results indicate that the basic emissions, transport, and transformation processes affecting summertime ozone levels in the Los Angeles basin are treated adequately.

Further development and testing work needed on the 3AM model includes:

- Implementation and testing of the secondary nitrate, sulfate, ammonium and organic aerosol chemistry
- Diagnostic testing of individual model algorithms including wind fields, mixing heights, wet and dry deposition, and chemical transformation
- Development of an appropriate methodology for developing weekend, holiday, and seasonal emissions inventory scaling factors
- Performance testing of the complete 3AM modeling system with photochemical and PM-10 air quality data bases

ACKNOWLEDGMENTS

This research was partially supported by a grant from the National Science Foundation and utilized the CRAY X-MP48 system at the National Center for Supercomputing Applications, University of Illinois at Urbana-Champaign.

REFERENCES

1. Cole, H. S., C. F. Newberry, W. Cox, G. K. Moss, and D. Layland. "Application of the Airshed Model for Ozone Control in St. Louis", 75th Annual Meeting of the Air Pollution Control Association, New Orleans, LA, June 20 to 25, 1982.
2. Schere, K. L., and J. H. Schreffler. "Final Evaluation of Urban-Scale Photochemical Air Quality Simulation Models," Draft Final Report, U.S. Environmental Protection Agency, Research Triangle Park, NC, 1982.
3. Gray, H. A. "Annual PM-10 Dispersion Model Development and Application in the South Coast Air Basin," Draft Air Quality Management Plan, 1988 Revision, Appendix V-L, South Coast Air Quality Management District, El Monte, CA, 1988.

4. Tesche, T. W., D. E. NcNally, L. O. Edwards, J. W. Embree, J. A. Rogalla, E. F. Aul, and B. A. Buzun. "Evaluation of the Technical Basis for PM-10 Control Requirements in the South Coast Air Basin," prepared for the Western States Petroleum Association by Radian Corporation, Sacramento, CA, 1989.
5. Lamb, R. G. "Simulated Effects of Hydrocarbon Emissions Controls on Seasonal Ozone Levels in Northeastern United States: A Preliminary Study," prepared by the Atmospheric Sciences Research Laboratory, U.S. Environmental Protection Agency, Research Triangle Park, NC, 1988.
6. Dennis, R. L. "Aggregation Workshop Report," Draft Report of Workshop Held July 20 to 21 at Bolton Landing, NY, U.S. Environmental Protection Agency, Research Triangle Park, NC, 1988.
7. Renne, D. "RADM Aggregation Workshop Report," prepared by the Atmospheric Sciences Research Laboratory, U.S. Environmental Protection Agency, Research Triangle Park, NC, 1988.
8. Seinfeld, J. H. *Atmospheric Chemistry and Physics of Air Pollution* (New York: John Wiley & Sons, Inc., 1986).
9. Roth, P. M.. "History of Ozone Modeling," presentation at the Conference on Photochemical Modeling as a Tool For Decision-Makers," co-sponsored by the California Institute of Technology and the California Air Resources Board, Pasadena, CA, February 1 to 3, 1988.
10. Reynolds, S. D., T. W. Tesche, and L. E. Reid. "Introduction to the Urban Airshed Model and Its Usage" (EF78-53R4), report prepared for the U.S. Environmental Protection Agency, Office of Air Quality Planning and Standards, by Systems Applications, Inc., San Rafael, CA, 1979.
11. Tesche, T. W. "Accuracy of Ozone Air Quality Models," *J. Environ. Eng.* 114 (4):739–752 (1988).
12. Tesche, T. W., and D. E. McNally. "Three Dimensional Snow Transport Modeling Over Mountainous Terrain," International Snow Science Workshop, Whistler, BC, Canada, October 12 to 15, 1989.
13. Goodin, W. R., G. J. McRae, and J. H. Seinfeld. "An Objective Analysis Technique for Constructing Three-Dimensional Urban-Scale Wind Fields," *J. Appl. Meteorol.* 19:98–108 (1980).
14. Goodin, W. R., G. J. McRae, and J. H. Seinfeld. "Comparison of Interpolation Methods for Sparse Data: Application to Wind and Concentration Fields," *J. Appl. Meteorol.* 18:761 (1979).
15. Tesche, T. W. and M. A. Yocke. "Numerical Modeling of Wind Fields Over Mountainous Regions in California," *Conference on Sierra Nevada Meteorology*, American Meteorological Society, South Lake Tahoe, CA, June 19 to 21, 1979.
16. Pielke, R. A. and P. Mehring. "Use of Mesoscale Climatology in Mountainous Terrain to Improve Spatial Representativeness of Mean Monthly Temperatures," *Mon. Weather Rev.* 105–108 (1977).
17. Tesche, T. W., T. C. Meyers, and G. E. Moore. "Use of Dynamic Meteorological Models as Input to Photochemical Dispersion Models," 79th Annual Meeting of the Air Pollution Control Association, Minneapolis, MN, June 22 to 27, 1986.
18. Tesche, T. W. and M. J. Hillyer. "Statistical Analysis and Graphical Presentation of Ambient Weather Conditions Within SCE's Service Territory," No. 81162, report to the Southern California Edison Company by Systems Applications, Inc., San Rafael, CA, 1981.
19. Steyn, D. G. and T. R. Oke. "The Depth of the Daytime Mixed Layer at Two

Coastal Sites: A Model and Its Validation," *Boundary Layer Meteorol.* 24:161–180 (1982).

20. van Ulden, A. P. and A. A. M. Holtslag. "Estimation of Atmospheric Boundary Layer Parameters for Diffusion Applications," *J. Appl. Climate Meteorol.* 24: 1196 (1985).
21. Wheeler, N. J. "Development of Mixing Depth Fields for Photochemical Grid Modeling for the South Central Coast Air Basin of California," 82nd Annual Meeting of the Air And Waste Management Association, Anaheim, CA, June 25 to 30, 1989.
22. Scott, B. C. "Theoretical Estimates for Scavenging Coefficient for Soluble Aerosol as a Function of Precipitation Type, Rate, and Altitude," *Atmos. Environ.* 16:1753–1762 (1982).
23. Holtslag, A. A. M. and A. D. van Ulden. "Simple Scheme for Daytime Estimates of Surface Fluxes from Routine Weather Data," *J. Climate Appl. Meteorol.* 22:517–529 (1983).
24. Wilczak, J. M. and M. S. Phillips. "Indirect Estimation of Convective Boundary Layer Structure for Use in Pollution Dispersion Models," *J. Climate Appl. Meteorol.* 25:1609–1634 (1986).
25. "Procedures for Preparation of Emission Inventories for Volatile Organic Compounds. II: Emission Inventory Requirements for Photochemical Air Quality Simulation Models," EPA-450/4-79-018, U.S. Environmental Protection Agency, Research Triangle Park, NC, 1979.
26. "Procedures for Preparation of Emission Inventories for Precursors of Ozone. I: Emission Inventory Requirements for Post-87 Ozone State Implementation Plans," EPA-450/4-88-019, U.S. Environmental Protection Agency, Research Triangle Park, NC, 1988.
27. Whitten, G. Z., H. Hogo, and J. P. Killus. "Carbon Bond Mechanism: A Condensed Kinetic Mechanism for Photochemical Smog," *Environ. Sci. Technol.* 14: 699 (1980).
28. Whitten, G. Z. and M. W. Gery. "Development of CBM-X Mechanisms for Urban and Regional AQSM's," U.S. Environmental Protection Agency, Research Triangle Park, NC, 1986.
29. Hudischewskyj, A. B. and C. Seigneur, "Mathematical Modeling of Chemistry and Physics of Aerosols in Plumes," *Atmos. Environ.* 23:967–983 (1989).
30. Lurmann, F. L., W. P. L. Carter, A. C. Lloyd, and L. A. Coyner. "Surrogate Species Reaction Mechanism for Urban-Scale Air Quality Simulation Models," EPA/600/3-87/014, U.S. Environmental Protection Agency, Research Triangle Park, NC, 1987.
31. Gery, M. W., G. Z. Whitten, and J. P. Killus. "Development and Testing of the CBM-IV for Urban and Regional Modeling," U.S. Environmental Protection Agency, Research Triangle Park, NC, 1988.
32. Lamb, R. G. "A Regional Scale (1000 km) Model of Photochemical Air Pollution. Part I: Theoretical Formulation," Environmental Sciences Research Laboratory, U.S. Environmental Protection Agency, Research Triangle Park, NC, 1982.
33. Hogo, H., C. Seigneur, and M. A. Yocke. "Technical Description of a Photochemical Air Quality Model with Extensions to Calculate Aerosol Dynamics and Visibility," report to the South Coast Air Quality Management District by Systems Applications, Inc., San Rafael, CA, 1985.

34. Pilinis, C. and J. H. Seinfeld. "Development and Evaluation of an Eulerian Photochemical Gas-Aerosol Model," *Atmos. Environ.* 22:1985 (1988).
35. Pilinis, C., J. H. Seinfeld, and C. Seigneur. "Mathematical Modeling of Dynamics of Multicomponent Atmospheric Aerosols," *Atmos. Environ.* 21(4):943–955 (1987).
36. Warren, D. R., and J. H. Seinfeld. "Simulation of Aerosol Size Distribution Evolution in Systems with Simultaneous Nucleation, Condensation, and Coagulation," *Aerosol Sci. Technol.* 4:31–43 (1985).
37. Seigneur, C., P. Saxena, M. A. Yocke, and T. W. Tesche. "Modeling Secondary Aerosol Formation in the California South Coast Air Basin – A Feasibility Study," prepared for the U.S. Environmental Protection Agency, Region IX by Systems Applications, Inc., San Rafael, CA, 1984.
38. Bassett, M. E., and J. H. Seinfeld. "Atmospheric Equilibrium Model of Sulfate and Nitrate Aerosol. II: Particle Size Analysis," *Atmos. Environ.* 18:1163–1170 (1984).
39. Bassett, M. E. and J. H. Seinfeld. "Atmospheric Equilibrium Model of Sulfate and Nitrate Aerosol," *Atmos. Environ.* 15:2395–2406 (1983).
40. Saxena, P., A. B. Hudischewskyj, C. Seigneur, and J. H. Seinfeld. "Comparative Study of Equilibrium Approaches to Chemical Characterization of Secondary Aerosols," *Atmos. Environ.* 20:1471–1483 (1986).
41. Grosjean, D. and J. H. Seinfeld. "Parameterization of the Formation Potential of Secondary Organic Aerosols," Draft Air Quality Management Plan, 1988 Revision, Appendix J-N, South Coast Air Quality Management District, El Monte, CA, 1988.
42. Tesche, T. W., J. L. Haney, and R. E. Morris. "Performance Evaluation of Four Grid-Based Dispersion Models in Complex Terrain," *Atmos. Environ.* 21(1):233–256 (1987).
43. Chamberlain, A. C. "Transport of Gases to and from Gases and Grass-like Surfaces," *Proc. Roy. Soc. A* 290:236–265 (1966).
44. Sheih, C. M., M. L. Wesely, and B. B. Hicks. "Estimated Dry Deposition Velocities Over Northeastern United States," *Atmos. Environ.* 13:1361–1368 (1979).
45. Wesely, M. L., and B. B. Hicks. "Some Factors That Affect Deposition Rates of Sulfur Dioxide and Similar Gases on Vegetation," *J. Air Pollut. Control Assoc.* 21:1110–1116 (1977).
46. Walcek, J. M., R. A. Brost, J. S. Chang, and M. L. Wesely. "SO_2, Sulfate, and HNO_3 Deposition Velocities Computed Using Regional Land use and Meteorological Data," *Atmos. Environ.* 20:949–964 (1986).
47. Killus, J. P. et al. "Continued Research in Mesoscale Air Pollution Simulation Modeling, Vol. V," Report No. 77-142, Systems Applications, Inc., San Rafael, CA, 1977.
48. Chang, J. S., R. A. Brost, I. S. A. Isaksen, S. Madronich, P. Middleton, W. R. Stockwell, and C. J. Walcek. "A Three-Dimensional Eulerian Acid Deposition Model: Physical Concepts and Formulation, *J. Geophys. Res.* 92:14,681–14,700 (1987).
49. Wesely, M. L. "Parameterization of Surface Resistances to Gaseous Dry Deposition in Regional-Scale Numerical Models," *Atmos. Environ.* 23(6):1293–1304 (1989).
50. Hales, J. M.. "Parameterization of Removal Mechanisms," in *Modeling of the Urban Boundary Layer*, American Meteorological Society, Boston, MA.

51. Yanenko, N. N.. *Method of Fractional Steps* (New York: Springer-Verlag, Inc., 1971).
52. Chock, D. P. and A. M. Dunker. "Comparison of Numerical Methods for Solving the Advection Equation," *Atmos. Environ.* 17:11–24 (1983).
53. Hesstvedt, E. et al. "Quasi-steady-state Approximations in Air Pollution Modeling: Comparison of Two Numerical Schemes for Oxidant Prediction," *Int. J. Chem. Kinet.* 10:971–994 (1978).
54. Hov, O. et al. "Comparison of Numerical Techniques for Use in Air Pollution Models with Non-linear Chemical Reactions," *Atmos. Environ.* 23(5):967–983 (1989).
55. McRae, G. J., W. R. Goodin, and J. H. Seinfeld. "Numerical Solution of the Atmospheric Diffusion Equation for Chemically-Reacting Flows," *J. Comput. Phys.* 45 (1):1–42 (1982).
56. Schere, K. L. "Evaluation of Several Numerical Advection Schemes," *Atmos. Environ.* 17(10):1987–1997 (1983).
57. Smolarkiewicz, P. K. "Simple Positive Definite Advection Scheme with Small Implicit Diffusion," *Mon. Weather Rev.* 111:479–486 (1983).
58. Tesche, T. W. "Comparison of Two Numerical Schemes for Integrating the Atmospheric Diffusion Equation: Evaluation with Atmospheric Data," *J. Appl. Meteorol.* 9(7):507–519 (1987).
59. Hov, O. "Numerical Solution of a Simplified Form of the Diffusion for Chemically-Reactive Atmospheric Species," *Atmos. Environ.* 17(3):551–562 (1983).

CHAPTER 6

Tropospheric Chemistry of Natural Hydrocarbons and Energy-Related Pollutants: Biosphere/Climate Feedbacks?

Jeffrey S. Gaffney and Nancy A. Marley

INTRODUCTION

It has become increasingly apparent that atmospheric pollution is causing severe environmental effects on urban, regional, and global scales. Attempts to assess these problems and develop solutions have depended heavily upon computer modeling techniques. Physicochemical computer modeling has been a key tool in attempting to evaluate pollution control strategies, particularly for urban ozone control.[1] Unfortunately, due to the inherent complexities involved, many proposed solutions to atmospheric pollution problems ranging from urban to global have not taken into account the coupled consequences. Thus, use of tall stacks in the Ohio River Valley to solve local pollution problems due to sulfur dioxide (SO_2) emissions contributes to regional-scale acid rain deposition. Both temporal and spatial interactions need to be considered when examining air pollution control strategies.

We are becoming more aware of the potential interplay between energy-derived pollutants from the combustion of fossil fuels and the natural emissions of trace gases into the troposphere. This chapter will begin to outline some of the coupled physical and chemical processes that will affect future modeling and mitigation strategies for a variety of atmospheric pollution problems. Our air quality is a global concern that will require us to integrate a variety of issues to develop rational approaches. There is no panacea, and the only short-term solutions that we should consider are those that do not lead to long-term problems or to costly, ineffective controls. The use of modeling to evaluate "what if" scenarios for different environmental control strategies is recommended to examine the possible feedbacks between various problems. These sensitivity tests would range from simple "Gedanken experiments" to detailed physical and chemical computerized modeling exercises.

SOME EXAMPLES OF THE POTENTIAL PROBLEMS

Some of the potential pitfalls associated with short-term solution are outlined in the following pages. These problems are meant to be examples and not a comprehensive list.

Urban Ozone and Carbon Monoxide Control

Photochemical oxidant (i.e., ozone) formation has been identified as an important health problem. As well, elevated levels of carbon monoxide due to motor vehicular and stationary source emissions have been identified as a cause for concern. Numerous control strategies have been proposed, many of which are based upon "smog chamber" and simplified chemical modeling studies.[1] Proposed solutions have varied from the use of alternative fuels (e.g., alcohols) to the prohibition of private burning of wood and other combustible materials.

The main schools of thought have been to use substitute fuels which lead to lower primary emissions of carbon monoxide, as well as lower reactive hydrocarbon levels. The reactive hydrocarbons lead to the production of ozone via secondary photochemical oxidation reactions. These reactions have been outlined in detail previously by numerous researchers.[1] The key reactions are outlined briefly below starting with the photolysis of nitrogen dioxide (NO_2):

$$NO_2 + hv \rightarrow NO + O \qquad (1)$$

$$O + O_2 \rightarrow O_3 \qquad (2)$$

$$O_3 + NO \rightarrow NO_2 + O_2 \qquad (3)$$

These first three reactions lead to the so-called "photostationary state" production of ozone which will be dependent upon the levels of NO, NO_2, and light. Increased formation of ozone depends upon the conversion of NO to NO_2 without consumption of ozone. This is accomplished by the production of peroxy radicals from organics due to tropospheric oxidation, principally by hydroxyl radical. The following chemical transformation steps illustrate the important chain reactions:

$$O_3 + hv \rightarrow O_2 + O(^1D) \qquad (4)$$

$$O(^1D) + H_2O \rightarrow 2OH \qquad (5)$$

$$OH + CO \text{ or organics} \rightarrow HO_2,\ RO_2,\ \text{and/or}\ RCO_3 \qquad (6)$$

$$HO_2,\ RO_2,\ \text{and/or}\ RCO_3 + NO \rightarrow NO_2 + OH,\ RO,\ \text{and/or}\ RO_2 \qquad (7)$$

Thus, the conversion of NO to NO_2 by the peroxy radicals leads to the further production of the active oxidant, OH, and a chain reaction mechanism for increased ozone production. Key termination reactions are as follows:

$$OH + NO_2 \rightarrow HNO_3 \tag{8}$$

$$HO_2 + HO_2 \rightarrow H_2O_2 + O_2 \tag{9}$$

From these reaction sequences it can readily be seen that the formation of ozone can be controlled by two means, NO_x control or reactive hydrocarbon control.[2] In the internal combustion engine NO_x control is difficult at best. Therefore, most control strategies have attempted to minimize the primary emissions of reactive hydrocarbons into the atmosphere usually by the use of catalytic converters. Another approach has been to use alternative fuels that are lower in chemical reactivity, with the idea that the lower reactivity fuel will lead to slower production of ozone.

A number of problems have arisen from this approach. These include, but are not limited to, the fact that the alternative fuels may be broken down during combustion to produce reactive emission products, which can also be toxic. If these products are not currently designated as criteria pollutants, they may not be examined or considered in assessing the total impact on our air quality. The natural reactive hydrocarbon backgrounds and chemical fluxes may also be substantial and can act to override the anthropogenic hydrocarbon control strategies attempting to control the conversion of NO to NO_2. Although these control strategies may or may not lead to improving ozone concentration levels on urban and regional scales, they may lead to increases in other oxidants such as peroxyacetyl nitrate (PAN) or hydrogen peroxide or to increased levels of nitric acid.[2,3] This principally arises from the fact that higher ozone levels will lead to higher OH, and as NO is reacted away, the radical termination reactions such as Reactions 8 and 9 above will begin to increase in importance. Two specific examples of these potential problems are the use of alcohol fuels as alternative energy sources and the interaction of naturally emitted isoprene with combustion-produced NO_x emissions.

The use of ethanol in Brazil has been shown to lead to increases in PAN via increased primary emissions of acetaldehyde and formaldehyde from the combustion of gasohol mixtures.[4] The further reaction of aldehydes via photolysis and reaction with OH to form peroxy radicals has been well established. Thus, modeling studies have predicted that the use of these fuels may lead to increased aldehyde, PAN, hydrogen peroxide, and organic oxidant levels downwind of an urban center. Methanol/gasoline mixtures and methanol have also been proposed as clean fuels that will lead to improved air quality. Formaldehyde has long been known to be an important emission from methanol-fueled vehicles, with concentration levels typically five times greater than conventional gasoline powered engines. Additionally, preliminary measurements of methanol-fueled vehicular hydrocarbon emissions indicate that substantial levels of aromatics are emitted from older vehicles, most likely caused by motor lubricant cracking.[5] Formaldehyde is a photochemically active gas that will lead to the direct production of hydroperoxyl radicals. As pointed out previously, these radicals can oxidize NO leading to the catalytic formation of

Table 1. Atmospheric Lifetimes Expected for Isoprene and the Monoterpene Hydrocarbons α- and β-pinene for OH and Ozone Loss[1]

	half-life (hr)		
	Isoprene	α-Pinene	β-Pinene
OH radical (molecules cm^{-3})			
5×10^5	6	10	7
1×10^6	3	5	3.5
3×10^6	1	1.7	1.2
Ozone (ppb)			
30	26	4	18
60	13	2	9
120	6.5	1	4.5

ozone or react with themselves to form hydrogen peroxide. Although methanol is quite unreactive, its combustion product, formaldehyde, is not. Thus, the use of methanol fuels may be less effective in reducing air pollution than expected due to aldehyde formation.

It has also been shown that isoprene chemical oxidation in the atmosphere will lead to the production of elevated levels of formaldehyde and hydrogen peroxide, as well as ozone, when these natural hydrocarbons interact with a polluted urban plume.[2,6,7] Thus, oxidation of isoprene is predominantly determined by its reaction with hydroxyl radical. Expected lifetimes for isoprene and α- and β-pinene are given in Table 1 for ozone and OH loss, respectively. The formation of formaldehyde from isoprene oxidation is well known, and elevated levels of formaldehyde have been measured in a deciduous forest.[8] Formaldehyde oxidation will in turn lead to hydroperoxyl radical formation, which will react either with NO to form nitrogen dioxide (and hence ozone) or with itself to form hydrogen peroxide. Hydrogen peroxide is an important factor in acid rain formation. This is due to the fact that hydrogen peroxide is highly water soluble and will react via an acid-catalyzed chemistry with dissolved SO_2 to form sulfuric acid in wet aerosols, clouds, fogs, and precipitation. This aqueous chemical reaction is as follows:

$$H_2O_2 + SO_2 + H_2O\ (\ell) \rightarrow 2H^+ + SO_4^{2-} + H_2O\ (\ell) \qquad (10)$$

Thus, the formation of sulfuric acid in "acid rain" is coupled to the oxidative chemical reactions that can also produce ozone and other oxidants. Regional and urban air pollution control strategies that do not take this coupled oxidation/sulfate formation chemistry into account could lead to slight reductions in ozone with substantial increases in hydrogen peroxide and other soluble gases which would in turn have regional-scale wet depositional impacts. These systems are coupled, and their interactions cannot be ignored when examining the total effectiveness of any specifically targeted control strategy.[2] To the best of our knowledge, these chemical interactions can be dealt with using the existing models with regard to the chemical implications; however, the physical effects of these processes need to be considered as well. For example, sulfate aerosols will act as cloud condensation nuclei and may lead to important climatic effects. The coupling of chemical and physical models to address these issues will be necessary to address these questions.

Acid Rain and Wet Deposition of Pollutants

As alluded to in the previous section, various atmospheric processes lead to the production of oxidants and acidic species, both inorganic and organic. Isolation of issues is another problem that can occur when addressing the various perceived air pollution concerns. Earlier we indicated how ozone control strategies aimed at urban pollution could lead to regional increases in the oxidation of SO_2, leading to greater wet deposition of acidic species. It is necessary to point out here that the converse could also occur and to realize that sulfuric and nitric acids are not the only water-soluble compounds that are being wet deposited on regional scales.

As an example, let us take a look again at Reaction 10. It has been shown that this reaction is quite rapid and if there is sufficient SO_2 in the air, the hydrogen peroxide in solution will be completely and rapidly titrated. Atmospheric measurements in some areas have found that there is an excess of hydrogen peroxide in solution after the available SO_2 has been reacted. Levels of excess hydrogen peroxide in summertime precipitation as high as 200 μM have been reported.[9] What if SO_2 emissions are reduced by half as has been recently proposed? The implication is that the hydrogen peroxide concentrations in the precipitation are likely to increase.

What might the consequences of such simple changes be that are not directly related to the "acid rain" issue? One possibility that needs to be addressed is the increase in wet deposition of hydrogen peroxide and other soluble species such as formaldehyde. Summertime formaldehyde levels as high as 60 μM have been reported in precipitation in the eastern U.S.[10] Although only a few studies have been carried out, hydrogen peroxide has been reported as causing damage to Norwegian spruce in laboratory studies.[11] Since both hydrogen peroxide and formaldehyde have acid-catalyzed chemistries, the question remains whether SO_2 controls may cause increases in other nonacidic species that have harmful effects in wet deposition.[2]

Another potential consequence of Reaction 10 is an increase or decrease in cloud condensation nuclei (CCN) which could affect the amount of precipitation, regional cloud cover, types of cloud cover and persistence, as well as the types of precipitation impacting an area.[12] Thus, narrowly focused control strategies for acid rain reduction could lead to change in the wet depositional patterns of a variety of other soluble and toxic pollutants. The chemical and physical interactions of changing aerosol levels may also have important feedbacks in climate and coupled cloud interactions as outlined in the next section.

Table 2. Examples of Identified Long-Lived and Reactive Greenhouse Trace Gases and Potential Classes of Organics and Aerosols that may be Important in Radiative Processes

Long-lived Trace Greenhouse Gases
Perchlorofluorocarbons (Freons)
Nitrous Oxide
Methane
Methylchloroform
Carbon monoxide
Reactive Trace Greenhouse Gases
Ozone
Peroxyacetyl nitrate (PAN)
Formaldehyde
Ammonia
Organics and Carbonaceous Species That May Be Important
Organic nitrates
Aromatics
Substituted aromatics
Ketones and aldehydes
Carbonaceous soot

Climate Change and Cloud Interactions

A great deal of attention has been given to the potential global climatic heating of the earth caused by greenhouse gases, particularly CO_2. Recently, a number of other trace greenhouse gases have been reported which also contribute to tropospheric heating. These include both reasonably long-lived gases and reactive species (Table 2). Many of these trace gases are more effective than CO_2 in heating the troposphere. This is because CO_2 is at a concentration level that is already saturating with regard to its ability to absorb infrared (IR) radiation. Another contributing factor is that many of the trace level species have absorptions in infrared (IR) windows where CO_2 and water vapor do not absorb. Many other potential greenhouse gases are known, particularly organic nitrates that are produced as minor side products in photochemical smog reactions. Since ozone, PAN, formaldehyde, carbon monoxide, and other gases from energy-related combustion processes are greenhouse gases as well as toxic species, these atmospheric components may have both climatic and toxic impacts.

When examining control strategies, modelers also need to keep in mind that many tropospheric greenhouse gases can play important roles in the stratosphere. Ozone holes in the antarctic and arctic regions have been identified, and the connection with chlorofluorocarbons has been made.[1,2] It is interesting to note that these same chlorofluorocarbons are active greenhouse gases. Similarly, methane is an active greenhouse gas which can act as a source of water vapor in the stratosphere, leading to polar stratospheric clouds and amplification of chlorine-catalytic destruction of ozone.[13,14] Nitrous oxide is also an important source of NO_x in the stratosphere, as well as an important trace greenhouse gas. Both nitrous oxide and methane are predominantly biogenic in source.

Counterplaying the greenhouse gases' absorption of IR radiation from the earth is the backscattering of incoming radiation by clouds. Due to their ability to backscatter the incoming solar radiation, clouds can play an important role in cooling the atmosphere. With the increase in greenhouse gas species due to combustion and agricultural practices has also gone a concurrent increase in primary and secondary aerosols. These aerosols act as CCN for cloud formation, as well as acting as light scatterers and absorbers. The question remains whether we have been masking the greenhouse heating effect by gases with tropospheric cooling due to increased aerosol levels. Control strategies for any of these problems need to consider the other issues as they are all coupled (Figure 1).

The Importance of the Biosphere in Atmospheric Processes

It has been known for some time that the biosphere plays an important, if not critical, role in maintaining the balance of our atmosphere. It is well established that the biosphere is active in maintaining many of the major reactive gases in the atmosphere such as oxygen, water vapor, and CO_2. We are just now beginning to appreciate that many trace level emissions from microbes, phytoplankton, and terrestrial plants may be playing equally important roles in moderating the radiative balance of the troposphere.[2,15,16]

Recently, we have discovered that methanogenic bacteria are responsible for most of the methane in our atmosphere. Methane is an important reactive hydrocarbon on global scales, is a greenhouse gas, and as indicated earlier, can act as a source of water vapor for the stratosphere. We have learned through

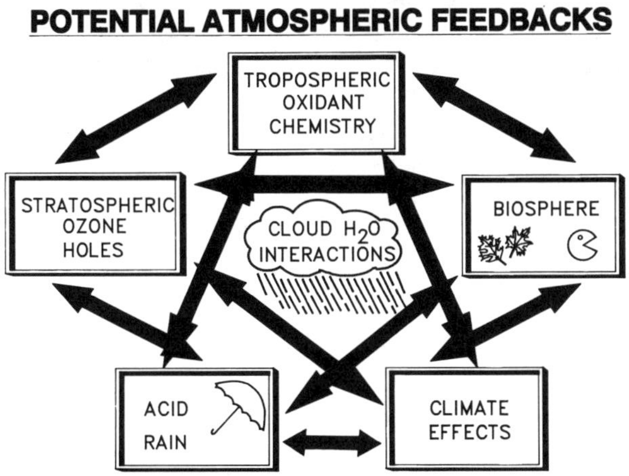

Figure 1. Diagram of identified environmental atmospheric issues, showing that these are not separate but interrelated problems (see text).

these studies that it is important to realize that various species of methanogens can produce a wide variety of emission rates for methane.

Similarly, research examining the importance of natural dimethylsulfide (DMS) in tropospheric processes has revealed that one phytoplankton species is the dominant source from marine systems. This gaseous emission is rapidly oxidized by OH to form methanesulfonic acid and ultimately sulfuric acid, which can act as CCN. Recently it has been proposed that this is a natural mechanism to increase cloud cover and precipitation.[15,16] Similar to the case of isoprene, the rate at which DMS is converted to CCN will be affected by the background ozone levels. Thus, under polluted situations, this rate will be accelerated and the spatial distribution of cloud formations will be affected.

As outlined in the previous section, emissions from forests can interact with energy-related NO_x to produce elevated levels of ozone and acidic sulfate. Another important biogenic gas is ammonia. This basic gas can react with either sulfuric or nitric acids to form ammonium salts. These can act as important aerosol species. Research on the "brown cloud" in Denver, CO has shown that the reaction of nitric acid with ammonia coming from feedlots nearby can lead to appreciable secondary aerosol.[19] This ammonium nitrate aerosol acts to efficiently scatter light and reduce visibility, thus contributing to the cloud.

When biogenic systems are considered in future modeling efforts, the emission of trace gases will need to be considered as a function of stresses on the biota, i.e., stresses due to air pollution, climatic change, and biospheric interactions. Thus, the emissions from plants are likely to change in composition and strength if the plant is or has been exposed to ozone, PAN, SO_2, peroxides, peracids, aldehydes, acids, etc. If drought or temperature extremes occur (i.e., frosts), these too will alter the biosphere/atmosphere exchange rates. Ecosystem interactions that may be synergistic with these effects would include parasitic infection, diseases, and insect infestation.

As we examine the potential strategies for urban ozone control, regional acid rain abatement, and reduction in global climatic impacts, we will need to determine these important biological/atmospheric feedbacks. The effects of varying temperature, relative humidity, and light intensity and wavelength will need to be factored into these assessments. The variability of emissions and susceptibility of numerous species of flora and fauna will have to be considered, as will their responses to varying precipitation, nutrient levels, and stresses.

CONCLUSIONS AND SUGGESTIONS FOR FUTURE MODELING EFFORTS

Hopefully, from the discussions presented, it is clear that the chemistry of the troposphere is a very complicated system with multiple feedbacks. If we are to come to grips with the problems presented by the combustion of fossil fuels, it will be important for us to better understand how the biosphere interacts with the atmosphere on urban, regional, and global scales. To accomplish this goal will require truly interdisciplinary study and model development.

One consideration that is becoming increasingly apparent is the fact that in the case of ozone production in the troposphere on urban and regional scales NO_x appears to be the limiting reagent.[1,2] Although recent studies have shown that natural NO_x due to lightning may be larger than anthropogenic sources from combustion of fossil fuels,[17] the distribution of lightning strikes is strongly dependent upon latitude. Thus, the relative input from lightning in the state of Florida is only a few percent of the total NO_x sources, compared to 60% due to lightning strikes on a global basis.[18] NO_x control strategies need to be more seriously explored, particularly in light of potential natural hydrocarbon interactions.

When one examines seriously the multiple environmental problems from the combustion of fossil fuels outlined here, it becomes increasingly apparent that we, as a global community, will need to examine alternate energy sources that do not pollute our air. Until these cleaner alternatives are implemented, we will need to design air quality control strategies that are effective and do not lead to unforeseen problems due to the types of interactions described in this chapter. Modeling can and should serve as an efficient and effective means of evaluating these potential interactions. These theoretical studies, along with field and laboratory research, will be needed to determine whether or not any one specific control methodology should be pursued.

ACKNOWLEDGMENTS

The authors wish to gratefully acknowledge the support of the U.S. Department of Energy Office of Health and Environmental Research Division of Atmospheric and Climatic Research (Contract W-31-109-ENG-38). This research was performed at Argonne National Laboratory. We wish to thank Dr. R. M. Miller, Ms. J. Jastrow, and Mr. D. Sisterson for helpful discussions.

REFERENCES

1. Finlayson-Pitts, B. J., and J. N. Pitts, Jr. *Atmospheric Chemistry: Fundamentals and Experimental Techniques* (New York: John Wiley & Sons, 1986) and references therein.
2. Gaffney, J. S., G. E. Streit, W. D. Spall, and J. H. Hall. "Beyond Acid Rain: Do Soluble Oxidants and Organic Toxins Interact with SO_2 and NO_x to Increase Ecosystem Effects?," *Environ. Sci. Technol.* 21:519–524 (1987).
3. Gaffney, J. S., N. A. Marley, and E. W. Prestbo. "Peroxyacyl Nitrates (PANs): Their Physical and Chemical Properties," in *The Handbook of Environmental Chemistry, Vol. 4/Part B*, O. Hutzinger, Ed. (Berlin: Springer-Verlag, 1989), pp. 1–38.

4. Tanner, R. L., A. H. Miguel, J. B. de Andrade, J. S. Gaffney, and G. E. Streit. "Atmospheric Chemistry of Aldehydes: Enhanced Peroxyacetyl Nitrate Formation from Ethanol-Fueled Vehicular Emissions," *Environ. Sci. Technol.* 22:1026–1034 (1988).
5. Gaffney, J. S., N. A. Marley, P. Doskey, and K. Picel. Preliminary data, Argonne National Laboratory, Argonne, IL (1989).
6. Gaffney, J. S. and G. I. Senum. "Peroxides, Peracids, Aldehydes, and PANs and Their Links to Natural and Anthropogenic Organic Sources," in *Gas-Liquid Chemistry of Natural Waters*, L. Newman, Ed. Brookhaven National Laboratory, Upton, NY, New York Report-BNL-51757, Vol. 1 (April 1984), pp. 5-1 to 5-7.
7. Chameides, W. L., R. W. Lindsay, J. Richardson, and C.S. Kiang. "The Role of Biogenic Hydrocarbons in Urban Photochemical Smog: Atlanta as a Case Study," *Science* 241:1473–1475 (1988).
8. Tanner, R. L. and J. S. Gaffney. "Biogenic and Anthropogenic Carbonaceous Aerosols and Their Precursors Within a Forest Canopy, Air Pollution Control Precursors Within a Forest Canopy," Air Pollution Control Association (APCA), San Francisco, CA, Paper No. 84-16-2 (1984).
9. Kelly, T. J., P. H. H. Daum, and S. E. Schwartz. "Measurement of Peroxides in Cloudwater and Rain," *J. Geophys. Res.* 90:7861–7871 (1985).
10. Lee, Y.-N., J. Shen, and P. Klotz. "Chemical Composition of Precipitation at Long Island, NY," *Water Air Soil Pollut.* 30:143–152 (1986).
11. Masuch, G., A. Kettrup, R. K. A. M. Mallant, and J. Slanina. "Effects of H_2O_2 Containing Acid Fog on Young Trees," *Anal. Environ. Chem.* 27(3):180–213 (1986).
12. Schwartz, S. E. "Control of Global Cloud Albedo and Climate by Marine Dimethylsulfide Emissions: A Test by Anthropogenic Sulfur Dioxide Emissions" *American Meteorological Society, Preprint Volume*, Anaheim, CA, January 30 to February 4, 1989, pp. 4–8.
13. Senum, G. I. and J. S. Gaffney. "A Reexamination of the Tropospheric Methane Cycle: Geophysical Implications," in *The Carbon Cycle and Atmospheric CO_2: Natural Variations Archean CO_2: Natural Variations Archean to Present* (Washington, D.C., American Geophysical Union, 1985), Geophysical Monogr. 32, pp. 61–69.
14. McElroy, M. B. and R. J. Salawitch. "Changing Composition of the Global Stratosphere," *Science* 243:763–770 (1989).
15. Charlson, R. J., J. E. Lovelock, M. O. Andreae, and S. G. Warren. "Oceanic Phytoplankton, Atmospheric Sulphur, Cloud Albedo and Climate," *Science* 326:655–661 (1987).
16. Schwartz, S. E. "Are Global Cloud Albedo and Climate Controlled by Marine Phytoplankton?," *Nature* 336:441–445 (1988).
17. Franzblau, E. and C. J. J. Popp. "Nitrogen Oxides Produced from Lightning," *J. Geophys. Res.* 94:11,089–11,104 (1989).
18. Sisterson, D. L. Personal communication (1989).
19. Calvert, J. G. Personal communication (1990).

CHAPTER 7

Regional Acid Deposition Modeling

Akula Venkatram, Gürkan Kuntasal, and Prakash Karamchandani

INTRODUCTION

Understanding of the acid deposition system implies achieving three overlapping objectives, which are to (1) characterize the relationships among observed variables, (2) characterize the relationships among the governing processes, and (3) understand how the processes work. Models have been developed to achieve one or more of these objectives. The empirical model focuses on relationships among observations. The semi-empirical model attempts to fulfill the first two objectives, while the comprehensive model attempts to achieve all three objectives. This paper examines the development and application of semi-empirical and comprehensive models. A detailed discussion of acid deposition models is out of the scope of this paper because of the wide variety of topics that need to be covered in such a discussion. We can only provide an overview of typical long-range transport models and their applications to determine source-receptor relationships. The reader is encouraged to refer to the appropriate literature for more details.[1]

In developing a semi-empirical model, one attempts to be as realistic as possible in modeling the *relationships* among processes. However, the processes themselves are highly parameterized; simple functions or even parameters are used to represent processes. These parameterizations cannot be generally defended on purely scientific grounds. However, their simplicity provides an important advantage, which is that they can be adjusted or modified to provide an "optimum" fit between model predictions and observations. This leads to the charge that is often leveled against semi-empirical models: because they are tuned to a particular set of observations, they cannot be used to examine situations that are very different from those used to derive their parameterizations. This charge is valid to a degree. However, if the framework of the model is physically realistic, it is often possible to use our prior understanding of the processes to estimate the range of applicability of the derived parameterizations.

The approach governing the formulation of semi-empirical models should be contrasted with that used in constructing so-called comprehensive models. In comprehensive modeling, the system being studied is reduced to a number of components. Detailed submodels are then constructed for each of these component processes by using information derived from the individual processes from both the field and the laboratory. These submodels incorporate formulations that can be modified to improve the fit between model predictions and observations. The modification usually consists of a reformulation of the submodel, where the reformulation itself must be consistent with our understanding of the relevant process. In the semi-empirical model, this sort of modification is more on the "surface" of the submodel, whose deliberate simplicity implies lack of depth. The final parameterization in a semi-empirical model represents the effective role of the relevant process in the integrated behavior of the modeled system. This less-than-satisfactory understanding of the processes is compensated to a certain extent by the usually better description of observations that the semi-empirical model provides.

Thus, comprehensive and semi-empirical models play complementary roles in the modeling process. The comprehensive model emphasizes understanding of the mechanisms that underlie processes. The description of observations is also an important objective. However, the complexity of the model, coupled with inevitable errors in the large data sets required by comprehensive models, makes the goal difficult to achieve. In the semi-empirical model, adequate description of the observations is the primary concern; thus, explicit fitting of model predictions to observations is an integral part of model development. If a comprehensive model is viewed as a dynamic filing system for our knowledge of the acid deposition system, it can provide the type of qualitative information required to improve semi-empirical models. Results from sensitivity tests on even a partially tested comprehensive model might suggest the range of applicability of the simplifications made in semi-empirical models.

REPRESENTATIVE SEMI-EMPIRICAL MODELS

The first semi-empirical long-range transport models for acid deposition were developed to explain the causes of acid rain observed in Scandinavia in the early 1960s. The model by Eliassen and Saltbones[2] developed in 1975 is typical of these models and will be discussed here. It is a simple Lagrangian model based on the evolution of the mass of material enclosed within a control volume. The authors used sulfur in rain as a surrogate for acidity. Thus, the model's governing equations are based on a mass balance for SO_2 and for sulfate, its oxidation product.

Eliassen and Saltbones[2] assumed that the wet and dry scavenging of sulfur was proportional to the concentration, where the proportionality constant was λ for SO_2 and $\bar{\lambda}$ for sulfate. Similarly, the conversion rate of SO_2 to sulfate was represented by a first-order rate constant, k. They assumed that the entrainment of material into the control volume could be neglected.

These assumptions allowed them to write the following simple expressions for the evolution of the concentrations (averaged over the control volume) of SO_2 (p) and sulfate (q).

$$\frac{dp}{dt} = \frac{Q(t)}{h} - \lambda p - kp \qquad (1)$$

$$\frac{dq}{dt} = -\bar{\lambda} q + kp \qquad (2)$$

In Equation 1, Q(t) is the SO_2 emission rate and h is the height of the control volume, which can be identified with the top of the planetary boundary (mixed) layer. It is thus implicit that most of the sulfur remains within the mixed layer.

Because of variations in wind speed and direction within the mixed layer, it is difficult to identify the velocity that should result in the advection of the control volume. Furthermore, because the control volume is an ill-defined quantity, the choice of advection velocity is somewhat arbitrary. Eliassen and Saltbones[2] used the winds on the 850-millibar constant-pressure surface with the justification that upper-level winds at this pressure level represented boundary layer averages.

To calculate the concentrations at a receptor, backward trajectories were calculated at specified intervals. These trajectories originated at the receptor and went back by 48 hr. Equations 1 and 2 were solved using these trajectories, and the daily averages of SO_2 and sulfate were obtained by calculating a weighted average of the concentrations associated with each of these trajectories. All calculations were performed during dry periods to avoid complications with modeling of wet scavenging.

The dry removal rate of sulfate, $\bar{\lambda}$, was neglected with the justification that measurements and theory indicated that it was much slower than the formation rate. Then, the only parameters of the model are λ, k, and h. These were estimated by fitting model predictions to daily averages of ambient SO_2 and sulfate concentrations measured at several stations in Europe. The fit was reasonable; correlation coefficients ranged from 0.4 to 0.7. The major results of this study were estimates of the parameters λ, k, and h. The dry scavenging rate, λ, turned out to be 10^{-5}/sec, k was found to be 1%/hr, and the mixed-layer height, h, was 1 km.

Eliassen and Saltbones[3] have improved their model by incorporating better descriptions of vertical mixing and wet and dry deposition. The model was recently modified to study nitrogen transport and chemistry.[4] The model is still very simple and is thus incapable of explaining more than 50% of the observed variance of daily averages of concentrations or depositions. The model performs much better in estimating seasonal and annual averages, presumably because the short-term fluctuations neglected by model parameterizations are averaged out for the longer time periods. This result has been borne out in several studies, and consequently most semi-empirical Lagrangian models have been used primarily to estimate seasonal and annual concentrations.

Bhumralkar et al.[5] were among the first to apply a Lagrangian model to estimate long-term concentrations in North America. Samson and his group at the University of Michigan used a similar model to examine source-receptor relationships in North America.[6] Other relevant studies are those of Ellenton et al.[7,8]

Bolin and Persson[9] showed that, for long-term applications, Lagrangian models could be simplified by treating vertical and horizontal dispersion independently. This represented the first step toward developing a truly statistical approach to long-range transport modeling. The basic assumption here is that processes such as scavenging and transport can be treated independently when we want to predict long-term values. This simplifies the framework of the model to the extent that we can usually fit model predictions to observations by adjusting the parameterizations of the processes. The validity of this approach is determined largely by the model's ability to explain observations adequately.

Sheih[10] followed up on the work of Bolin and Persson[9] by developing a similar model suitable for applications to North America. Shannon[11] developed an improved version of the model, which is referred to as the Advanced Statistical Trajectory Regional Air Pollution (ASTRAP) model. Results from some of the studies conducted with ASTRAP will be discussed here.

Fisher[12] simplified the statistical approach even further by incorporating straight line trajectories. He also adapted a stochastic scavenging model proposed by Rodhe and Grandell.[13] Although the model dealt only with SO_2, it provided reasonable estimates of long-term wet deposition of sulfur in Europe. Encouraged by this result, Venkatram et al.[14] extended Fisher's approach by modifying the stochastic scavenging model to account for sulfate as well as SO_2. Smith[15] developed a similar extension independently.

In most of the studies mentioned thus far, the adjustment of model parameters to fit model predictions to observations was only implicit and was usually lost in the process of model development. Venkatram and Pleim[16] were among the first investigators to make optimization of model parameters an explicit procedure that needed to be examined and described in its own right. As we will see later, the results of this optimization supplement the results on source-receptor relationships.

As mentioned earlier, semi-empirical models traditionally treat chemistry and scavenging using rate coefficients that do not depend on the concentration level. Although they have been relatively successful in explaining observations of wet sulfur deposition, our knowledge of the role of oxidant limitation in wet scavenging suggests that we must be careful about using their results to understand the effects of emission controls on wet deposition. Recent studies indicate that oxidant limitation can introduce nonlinearity close to large source areas.[17] However, the effect appears to be insignificant in the remote areas that are associated with sensitive receptors. This will be discussed in more detail in the next section.

Results on Nonlinearity

The effects of nonlinearity can be examined indirectly by looking at the performance of "linear" models in explaining observed patterns of wet deposition. The scatterplot of Figure 1 is typical of the performance of semi-empirical models.[16] It compares model predictions of annual wet deposition of sulfur for 1980 against observations made at 62 stations distributed over the northeastern U.S. and southeastern Canada. Although the scatter is not small, the model shows little bias. The r^2 (square of the correlation coefficient) is about 0.6—the model explains 60% of the observed variance. These results would suggest that there is little need to invoke nonlinearity. However, the effects of nonlinearity might be buried in the relatively large scatter between model predictions and observations. We need to reduce the scatter before we can come to definite conclusions. To see how this can be done, it is necessary to retrace the steps involved in using the semi-empirical model to understand wet deposition.

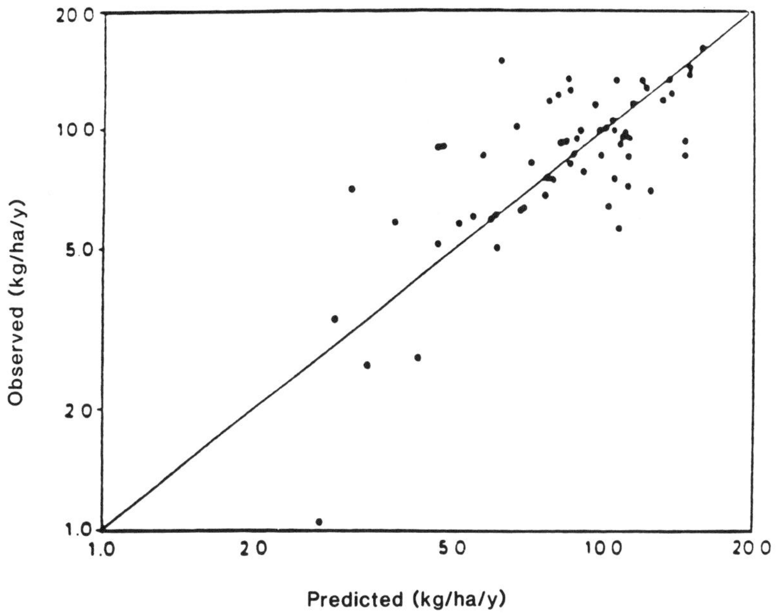

Figure 1. Comparison of model estimates with observations of annual wet sulfur deposition for 1980 at 62 stations in the eastern U.S. Background of 2 kg/ha/year has been added to model estimates. (Reprinted from Venkatram, A. and J. Pleim. "Analysis of Observations Relevant to Long-Range Transport and Deposition of Pollutants," *Atmos. Environ.* 19:659–667 (1985). Copyright 1985, Pergamon Press, Elmsford, N.Y. With permission.)

We indicated earlier that changes in wet deposition would be proportional to emission changes if sulfate produced in the gas phase dominated the sulfur in rain. To see if the data indicated that this was a possibility, Venkatram and Pleim[16] compared the observations mentioned in the previous paragraph to the predictions of a semi-empirical model in which the wet deposition of SO_2 was set equal to zero. Because dissolution of SO_2 in rain is negligible, this is equivalent to assuming that incloud oxidation of SO_2 is zero. The resulting comparison, shown in Figure 2, indicates clearly that incloud oxidation makes the dominant contribution to sulfur in rain measured over a year.

The above result also indicates that because incloud oxidation is important, nonlinearity is also potentially important. This was investigated by Venkatram et al.[17] in a study in which a semi-empirical model was explicitly fitted to wet sulfur deposition data collected during the period of 1982 to 1985 at 87 locations in the eastern U.S. and Canada. One new feature of the model was the incorporation of oxidant limitation in the treatment of wet scavenging. The wet scavenging rate during any one precipitation event was proportional to a fixed oxidant concentration when the SO_2 concentration exceeded the oxidant concentration; otherwise the linear rate proportional to the SO_2 concentration

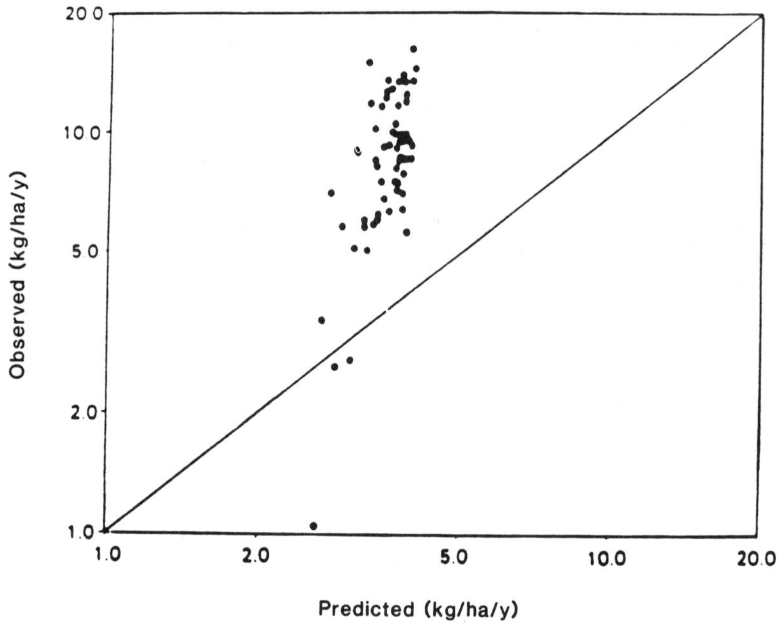

Figure 2. Comparison of model estimates with observations of annual wet sulfur deposition for 1980 at 62 stations in the eastern U.S. Background of 2 kg/ha/year has been added to model estimates and only sulfate formed during dry periods contributes to sulfur in rain. (Reprinted from Venkatram, A. and J. Pleim. "Analysis of Observations Relevant to Long-Range Transport and Deposition of Pollutants," *Atmos. Environ.* 19: 659–667 (1985). Copyright 1985, Pergamon Press, Elmsford, N.Y. With permission.)

was used. The oxidant concentration was treated as a parameter whose value was obtained by fitting model predictions to observations.

In the study, the values of six parameters—the mixed layer height, the average length of a dry period, the oxidant concentration, and three wind-direction frequencies—were optimized to minimize the sum of the squares of the differences between model predictions and observations at these stations. The values of the optimized parameters were consistent with the corresponding observations, indicating that there was no reason to seriously doubt the model's validity. Furthermore, the fit between model predictions and observations was excellent, as shown in Figure 3; the r^2 was 0.9. Thus, by explicit optimization of parameters, Venkatram et al.[17] were able to reduce the scatter between model predictions and observations.

The major result of this exercise was that the optimum oxidant concentration turned out to have a value equivalent to about 1 ppbv of H_2O_2. Neglecting oxidant limitation resulted in the fit shown in Figure 4, in which the model tends to overpredict sulfur concentrations in rain at the high end of the observations. This clearly demonstrates the effect of nonlinearity on long-term concentrations of sulfur in rain. However, in order to examine the

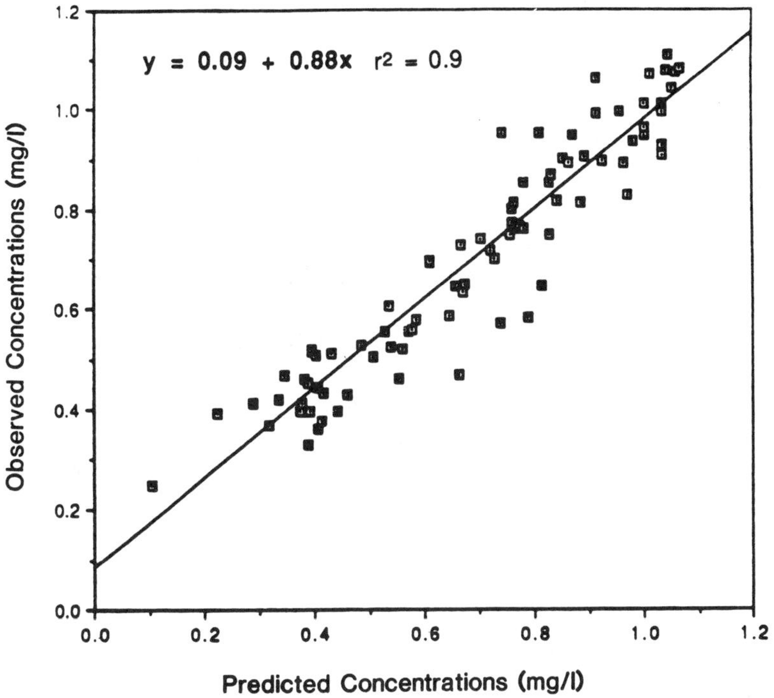

Figure 3. Comparison of nonlinear (oxidant limitation included) semi-empirical model predictions with observations of sulfur concentrations in rain averaged over 1982 to 1985. The best fit line is shown in the figure.

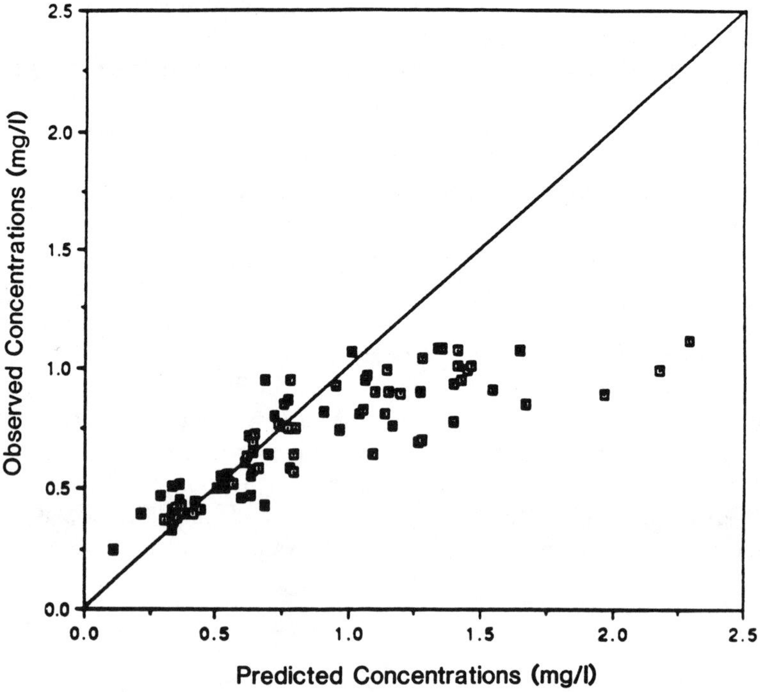

Figure 4. Comparison of linear (oxidant limitation neglected) semi-empirical model predictions with observations of sulfur concentrations in rain averaged over 1982 to 1985.

consequences of nonlinearity on deposition patterns, we need to use the semi-empirical model to simulate the effect of emission changes on predicted wet deposition patterns.

Figure 5 shows the results of a simulation in which SO_2 emissions across the grid were cut by half, and all other model inputs and parameters were held constant. Note that the background deposition was set to zero in both the base case and the emission reduction study. Thus, if wet deposition responded linearly, then the wet deposition across the grid would be reduced by 50%, and we would not see any gradients in the spatial pattern of wet deposition reductions. However, Figure 5 shows that there is a spatial variation in deposition reductions. The smallest reductions occur close to major source areas, as expected. For the space scale of the simulation, which is about 100 km, these reductions are about 35%, instead of the linear (or proportional) 50%. In regions several hundred kilometers from source regions, the reduction is close to the linear 50%.

Figure 6 shows the effect of adding a background deposition to the model prediction, which increases the apparent nonlinearity of the system. In remote areas, where background deposition makes a significant contribution, the

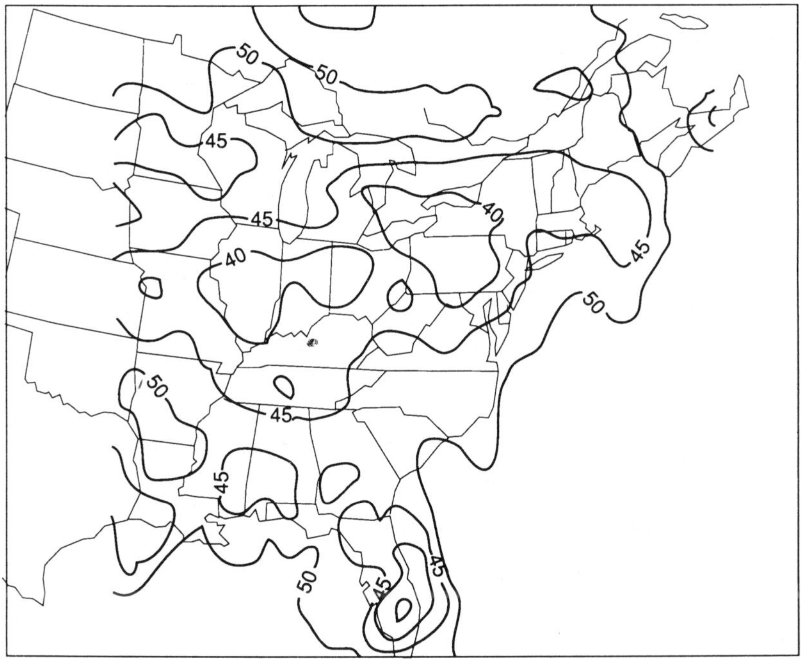

Figure 5. Relative decrease in sulfur wet deposition predicted by the nonlinear semi-empirical model when emissions across the grid are cut by 50%. No background deposition is added.

reduction in wet deposition is much less than the linear 50% because we have no control over the background deposition.

Simulations indicated that wet deposition of sulfur does react nonlinearly to SO_2 emission changes; however, the change in total deposition—wet plus dry—is almost linear. The reason for this is that where wet deposition is most nonlinear, it also contributes least to total deposition. This is clearly seen in Figure 7, which compares the predicted wet and dry deposition at the 87 locations mentioned earlier. Where wet deposition is limited by oxidant availability, dry deposition is much larger. Notice that while wet deposition does not exceed 13 kg/ha/year, dry deposition is as high as 30 kg/ha/year.

Results on Nitrogen Deposition

The relatively few attempts to model nitrogen deposition using semi-empirical models reflect the difficulty in formulating simple parameterizations for the chemistry responsible for converting NO_x to acidifying pollutants. Recall that the parameterization must account for the photochemistry that converts NO to NO_2, which in turn reacts with the OH radical to form nitric acid. Because both reactions affect the oxidant level, it is difficult to justify the use of a linear conversion rate. The nitric acid formed can exist in equilibrium

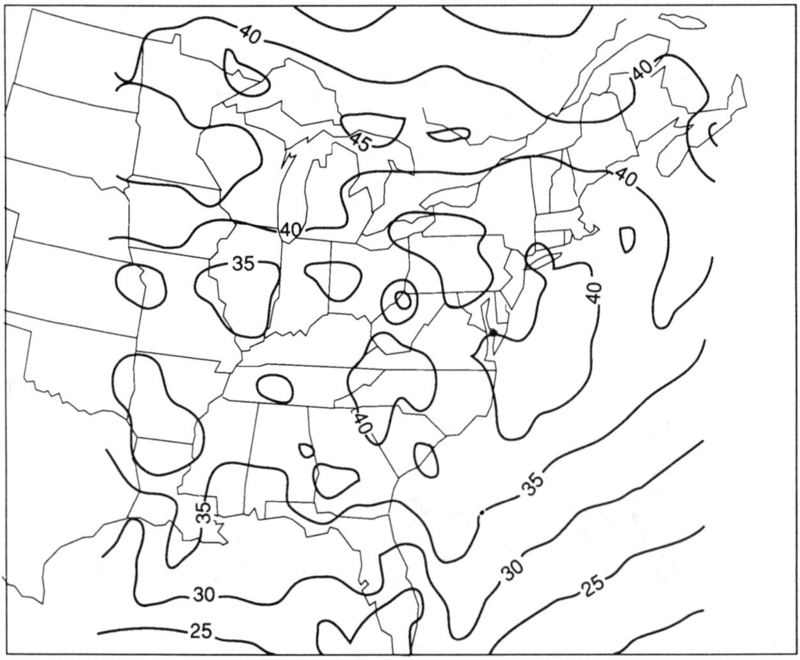

Figure 6. Relative decrease in sulfur wet deposition predicted by the nonlinear semi-empirical model when SO_2 emissions across the grid are cut by 50%. A background deposition corresponding to the regression equation of Figure 3 is used in the model calculations.

with inorganic nitrates in the form of submicron aerosols. We need to account for the partition between nitric acid and nitrate because of the large difference in the dry deposition velocities of the two species. Also, peroxyacetylnitrate, commonly referred to as PAN, needs to be treated explicitly because it can act as a mobile reservoir for NO_2 at low temperatures.

It would seem that it would be difficult to incorporate this chemistry in a semi-empirical model. Surprisingly, Shannon and Lesht[18] have obtained plausible results by assuming that the chemistry can be represented by a linear conversion of combined NO/NO_2 to an imprecisely defined nitrate. By judicious selection of the transformation rates and dry deposition velocities of the involved species, they obtained the fit shown in Figure 8. The model (ASTRAP) explains 74% of the variance of nitrate deposition measured in 1981. Part of this good correlation could be related to the use of the same variable—the observed precipitation—in both model predictions and observations (recall that wet deposition is the product of concentration in precipitation and the precipitation amount). Furthermore, because of the simplicity of the model parameterizations, it is difficult to determine whether the derived model parameters are consistent with our knowledge of the relevant processes.

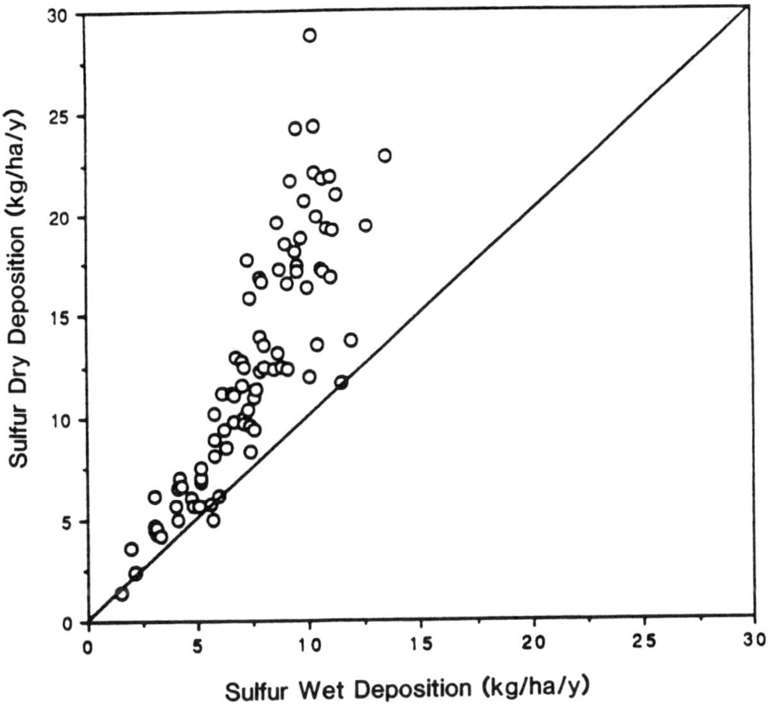

Figure 7. Comparison of predicted wet and dry deposition predicted by the nonlinear semi-empirical model at 87 locations in the eastern U.S. and Canada.

ASTRAP will have to be tested against several years of data before we can believe that its simple treatment of nitrogen chemistry is adequate.

Because the fit between model predictions and observations is good, it is worthwhile to examine the source-receptor relationships predicted by ASTRAP. Table 1 provides this information for nitrogen and sulfur, treating the U.S. and Canada as sources and receptors.[18] It is seen that U.S. sources are responsible for about 45% of the dry deposition of nitrogen in Canada and close to 90% of the wet deposition. This large difference between the relative contributions to dry and wet deposition in Canada could be related to the fact that NO_x must be converted to nitrate before it can be scavenged. Therefore, most of the wet deposited nitrogen in Canada consists of nitrate that originated as NO_x in the U.S.

The U.S. contributes about 65% to total nitrogen deposition in Canada. On the other hand, it is responsible for over 95% of its own deposition. This also applies to total sulfur deposition in the U.S. Furthermore, roughly half of the sulfur deposited in Canada originates from U.S. sources.

The treatment of sulfur chemistry and deposition relative to that of nitrogen is reflected in the differences in the contributions of the U.S. to the wet and dry depositions of sulfur and nitrogen in Canada. The U.S. makes a smaller

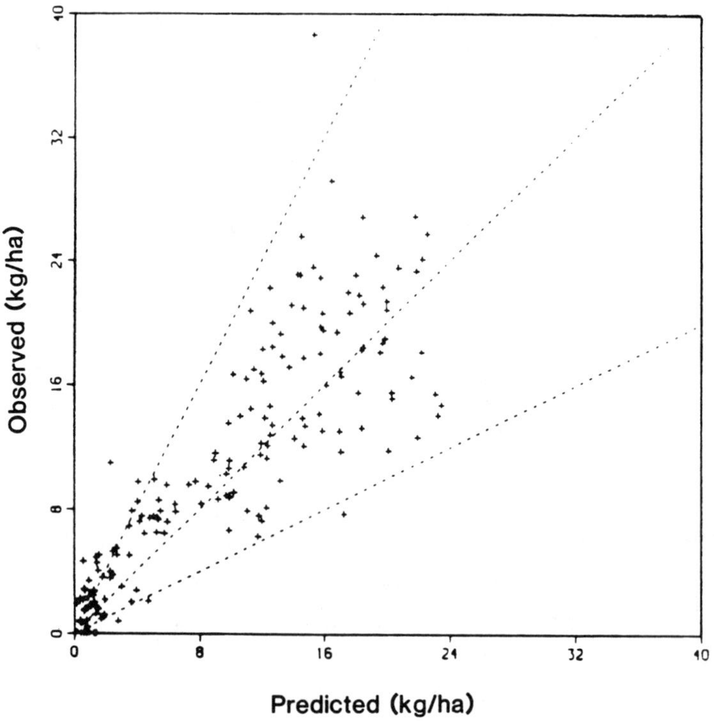

Figure 8. Comparison of ASTRAP simulations with observed annual wet nitrate deposition during 1981. (Dashed lines are for a perfect fit and for a factor of 2 range about a perfect fit.) (Reprinted by permission of Kluwer Academic Publishers from Shannon, J. D. and B. M. Lesht. "Estimation of Source-Receptor Matrices for Deposition of NO_x-N," *Water Air Soil Pollut.* 30:815–824 (1986).)

contribution to dry deposition of sulfur than to that of nitrogen because most of the sulfur reaching Canada from the U.S. is in the form of slowly depositing sulfate. Because ASTRAP assigns comparable dry deposition velocities to NO_x and nitrate, Canada makes a substantial contribution to its own dry deposition of nitrogen. Although sulfate from the U.S. is washed out efficiently over Canada, Canada contributes more to its sulfur wet deposition than to its nitrogen deposition because SO_2 originating from Canadian sources is also scavenged efficiently through incloud oxidation. On the other hand, NO_x from Canadian sources is not washed out until it is converted to nitrate, which happens when it travels to the U.S.

Another notable attempt to model nitrogen deposition is that of Brodzinsky et al.,[19] who incorporated a simple parameterization of nitrogen chemistry in the Eastern North American Model of Air Pollution (ENAMAP) Lagrangian puff model. Recognizing the need to distinguish between the different nitrogen species, they formulated their chemistry in terms of five species—NO, NO_2, HNO_3, PAN, and inorganic nitrate. These species were incorporated into five coupled linear differential equations describing their evolution in time. The

Table 1. Source-Receptor Matrix for Deposition in the U.S. and Canada

	Nitrogen					
Source/	Dry Deposition		Wet Deposition		Total Deposition	
Receptor	U.S.	Canada	U.S.	Canada	U.S.	Canada
U.S. (6,147 kT)	96	43	98	87	97	64
Canada (555 kT)	4	57	2	13	3	36
Deposition Magnitude (in kT)	1,771	254	1,360	394	3,131	648
	Sulfur					
Source/	Dry Deposition		Wet Deposition		Total Deposition	
Receptor	U.S.	Canada	U.S.	Canada	U.S.	Canada
U.S. (12,580 kT)	95	29	97	63	96	49
Canada (2,400 kT)	5	71	3	37	4	51
Deposition Magnitude (in kT)	3,490	841	2,702	1,079	6,192	1,920

Reprinted by permission of Kluwer Academic Publishers from Shannon, J. D. and B. M. Lesht. "Estimation of Source-Receptor Matrices for Deposition of No_x–N," *Water Air Soil Pollut.* 30: 815–824 (1986).

Note: Figures below source names are emissions from regions. Entries in the matrix are percentages; kT = kilotons.

choices of values for the rate constants (which included those for dry and wet deposition) appearing in these equations were guided by prior understanding of the relevant processes. The model has been used to predict concentrations and depositions of nitrogen species, although no attempt has been made to evaluate the predictions against observations. Because the model is semi-empirical, it must be calibrated against observations before it can be used to estimate source-receptor relationships.

COMPREHENSIVE MODELS

Two of the comprehensive models being used in North America are the Regional Acid Deposition Model (RADM), developed under funding from the U.S. Environmental Protection Agency (EPA), and the Acid Deposition and Oxidant Model (ADOM), developed under the sponsorship of Environment Canada (AES), Environment Ontario (OME), the West German Umweltbundesamt (UBA), and the Electric Power Research Institute (EPRI). Both models have been described elsewhere.[20,21] Here, we will confine our discussion to the results obtained from these models. It is important to point out that because these models are still in the first stages of testing, the results presented here should be considered preliminary and only broadly indicative of the behavior of the acid deposition system.

Results From ADOM and RADM

Both RADM and ADOM have been used to simulate wet deposition episodes during the Oxidation and Scavenging Characteristics of April Rains (OSCAR) field study conducted during April 1981. This study, which was designed to investigate wet removal during cyclonic storms, collected regional precipitation chemistry data in an intermediate-density network consisting of 37 stations in the eastern U.S. and Canada at a spatial resolution of 100 to 200 km. In addition to the regional data, intensive measurements were also made over a limited area using a high-density network consisting of 47 stations in a 110- by 110-km area in northeast Indiana/northwest Ohio. The details of the study are described in Easter et al.[22]

For the past 2 years, ADOM has been tested extensively against data from the OSCAR II (April 11 to 15, 1981) and OSCAR IV (April 21 to 24, 1981) episodes.[21,23,24] The actual testing involved a large number of simulations in which comparisons of model predictions with observations were used to diagnose and correct problems with the input data as well as the parameterizations in the process modules. This exercise showed that considerably more data than available from OSCAR would be required for even a marginally adequate evaluation of ADOM or RADM. The data collected by the Operational Evaluation Network (OEN) and ME-35 studies, sponsored by EPRI and EPA, respectively, might fill in part of the data gaps. These field studies, which are also referred to as the Eulerian Model Evaluation Field Study (EMEFS), were conducted in 1988 and 1989. The field studies provide a unique database of surface and aircraft measurements to evaluate regional scale acid deposition models such as ADOM and RADM. In addition to routine daily surface measurements of air quality (particulate sulfate, nitrate, ammonium, nitric acid, ammonia, sulfur dioxide, and nitrogen dioxide) and precipitation chemistry (rainfall amount, pH, major acidic and basic ions), surface and vertical measurements of a larger set of variables at finer temporal resolutions were collected during the field studies. These include measurements of SO_2, nitric acid, ozone, ammonia, NO_2, hydrogen peroxide, speciated hydrocarbons, particulate sulfate, nitrate, and ammonium, PAN, formaldehyde, and other aldehydes, as well as cloud and precipitation chemistry variables.

Figures 9 and 10 show the performance of the most recent version of ADOM in explaining event-averaged observations of sulfate and nitrate in rain made during the OSCAR field study.[24] We see that for OSCAR II, over 70% of the predictions of sulfate concentrations in rain are within a factor or two of the observations. The comparison for nitrate concentrations is not as good—only 60% of the predictions are within a factor of two of the observations. The

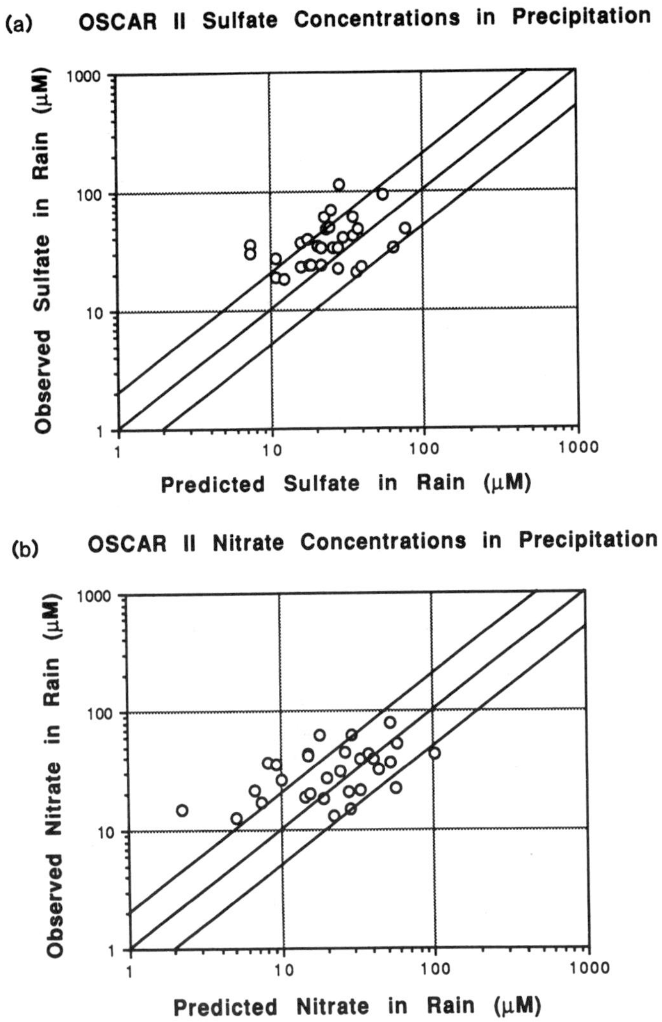

Figure 9. Comparison of event-averaged observed and ADOM-predicted (a) sulfate and (b) nitrate concentrations in precipitation for OSCAR II. Diagonal line indicates perfect agreement and the lines parallel to the diagonal line indicate agreement within a factor of two. (Courtesy of ENSR Consulting and Engineering, Camarillo, CA, "Final Report: Winter Episode Study," ENSR 5210-004-100 (1989).)

model provides a much better explanation for the observations made during OSCAR IV. Over 80% of the predictions of both sulfate and nitrate in rain are within a factor of two of the observations. However, the correlation coefficients (r) between observations and predictions of sulfate and nitrate in rain for the two episodes range from 0.3 to 0.5. On the other hand, the correlation coefficients between ADOM predictions and observations of wet deposition of sulfate and nitrate for OSCAR II and IV range from 0.6 to 0.7, presumably

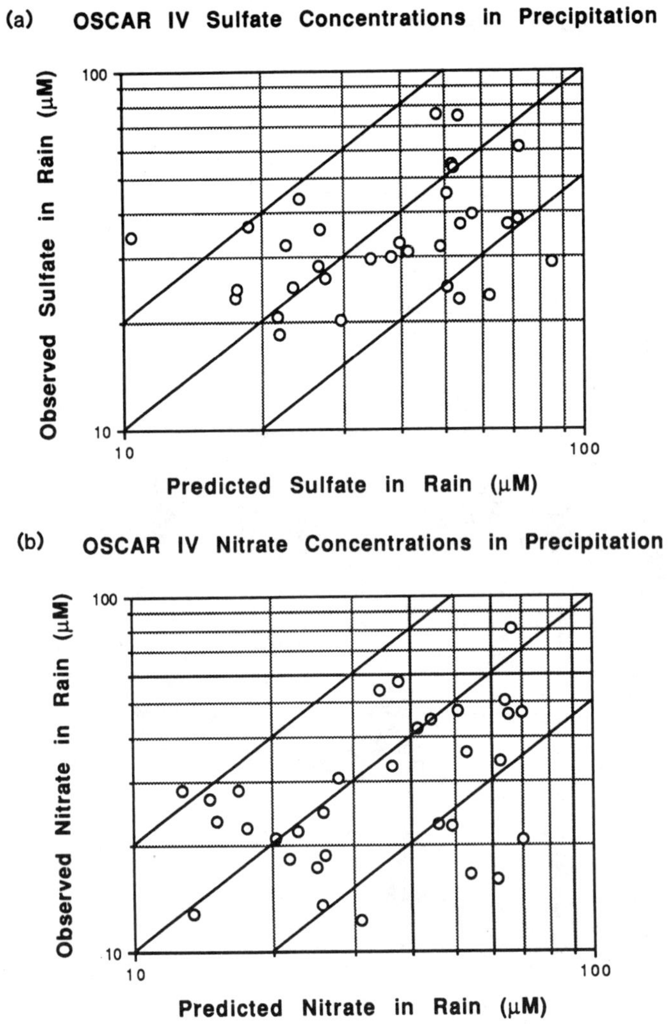

Figure 10. Comparison of event-averaged observed and ADOM-predicted (a) sulfate and (b) nitrate concentrations in precipitation for OSCAR IV. Diagonal line indicates perfect agreement and the lines parallel to the diagonal line indicate agreement within a factor of two. (Courtesy of ENSR Consulting and Engineering, Camarillo, CA, "Final Report: Winter Episode Study," ENSR 5210-004-100 (1989).)

due to the high degree of correlation between the precipitation amounts used in the simulations and observed precipitation amounts at the monitoring sites.

Figure 11 compares event-averaged observations of sulfate and nitrate wet deposition during OSCAR IV with predictions from RADM.[25] In general, the performance of RADM is similar to that of ADOM. RADM has also been exercised for OSCAR I and OSCAR II. However, like ADOM, the best RADM performance is for OSCAR IV—the data collected during OSCAR IV

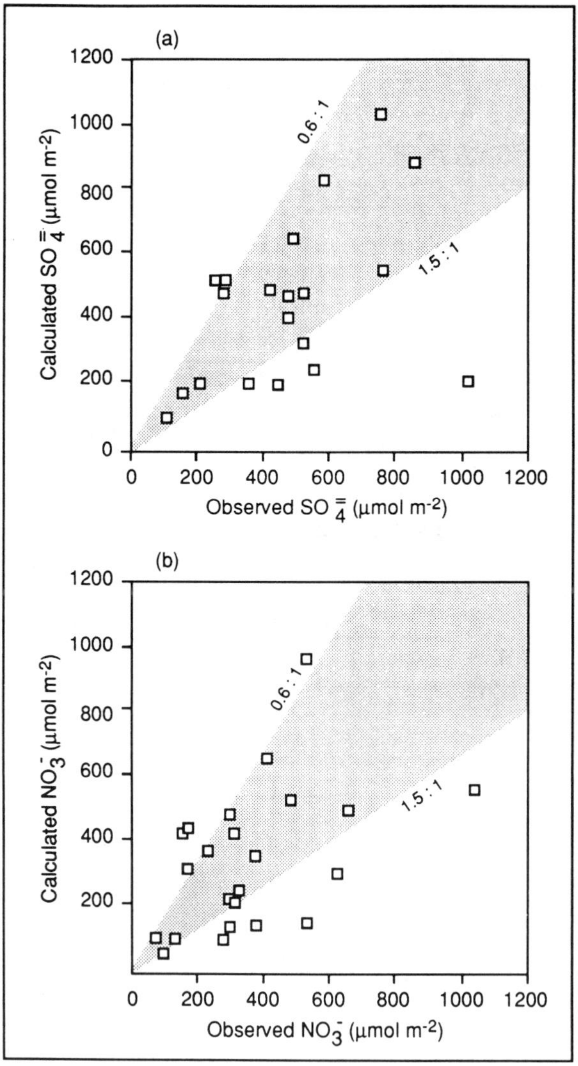

Figure 11. Comparison of event-total observed and RADM-predicted (a) sulfate and (b) nitrate deposition (OSCAR IV). (Reprinted from Walcek, C. J. and C. Berkowitz. "Modeling of Wet Scavenging and Subgrid-Scale Vertical Transport by Clouds in a Tropospheric Chemical Model," in *Air Pollution Modeling and Its Application VII*, H. van Dop, Ed. (New York: Plenum Publishing Corporation (1989), pp. 267–275. With permission.)

is considered to be of generally better quality than the data collected during the other OSCAR episodes.[26]

A part of the deviations between model predictions and observations might be related to the fact that model predictions represent grid averages while observations are essentially point values. We also have to contend with the inherent uncertainties caused by neglecting processes that are important on the time scale of a 5-day episode. Therefore, it should come as no surprise that the model predictions are associated with a 95% confidence interval of a factor of four when compared against point observations.

It is clear that we should not expect comprehensive models such as ADOM or RADM to provide numerically precise predictions. The main value of ADOM or RADM lies in their ability to yield an understanding of the source-receptor relationships that govern the acid deposition system. Before illustrating this feature, we make the point that both ADOM and RADM are comprehensive in the sense that they can simulate the photochemistry that underlies acid deposition. This is evident in Figure 12, which shows a comparison of a time series of ADOM predictions of ozone against corresponding observations made during the period of June 9 to 17, 1983.[27] Presumably because ozone is a secondary pollutant with large space scales, the model does a very creditable job of explaining observations.

Both ADOM and RADM have also been used to understand nonlinearities in acid deposition by investigating the impact of various emission control scenarios on acidic deposition. Figure 13 shows the percentage reduction in 20-day (April 10 to 29, 1981, corresponding to the OSCAR episode) averaged sulfate concentrations in rain predicted by ADOM when SO_2 and SO_4 emissions across the region shown in the figure are reduced by 50%.[28] The response of wet deposition is much less than linear—30 to 43%—in areas with high SO_2 emission densities. Note that the boundary conditions for this simulation were

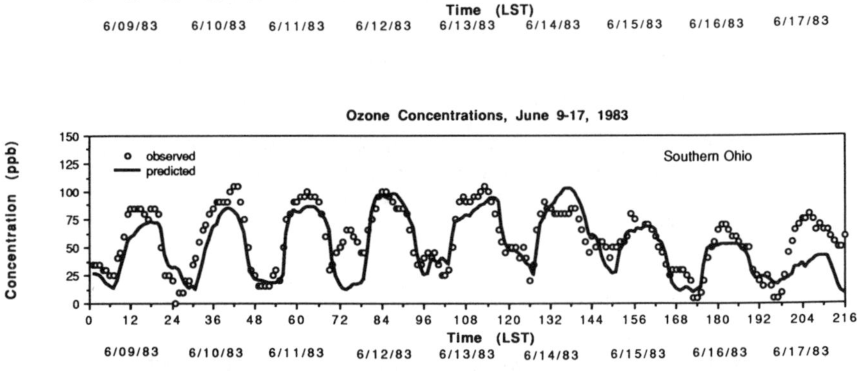

Figure 12. Time series of ADOM predictions and observations of ozone in northern Maryland and southern Ohio, for June 9 to 17, 1983.

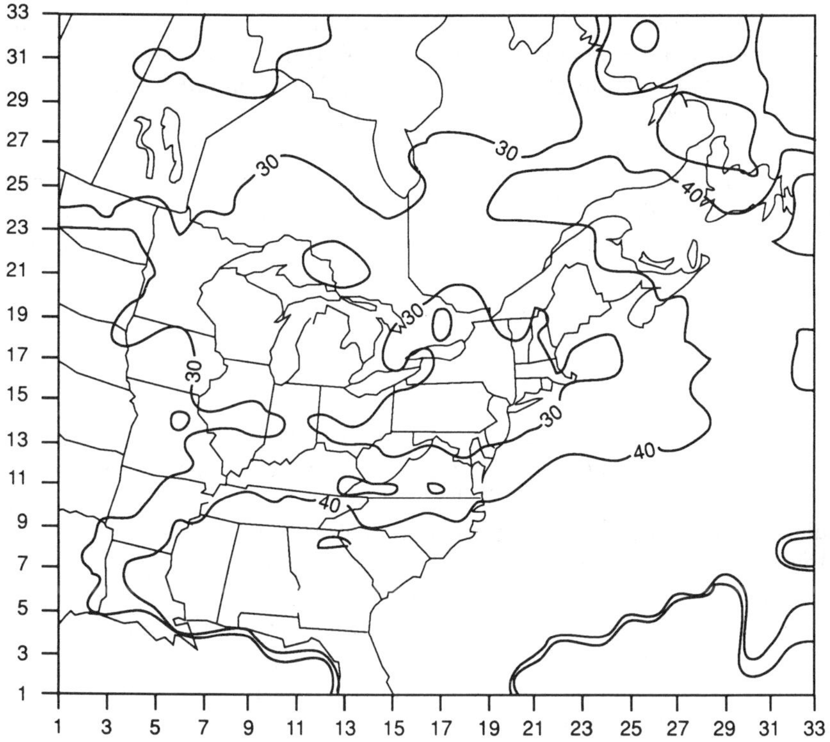

Figure 13. ADOM-predicted percentage reduction in 20-day (April 10 to 29, 1981) averaged SO_4^{2-} concentrations in precipitation as a result of a 50% across-the-grid reduction of SO_2 and sulfate emissions. SO_2 and sulfate boundary and initial conditions were not changed in the study. (Reprinted from Misra, P. K., R. Bloxam, C. Fung, and S. Wong. "Non-Linear Response of Wet Deposition to Emissions Reduction: A Model Study," *Atmos. Environ.* 23:671–687 (1989). Copyright 1989, Pergamon Press, Elmsford, N.Y. With permission.)

the same as the base case values. Although this assumption causes an artificial nonlinear response near the boundaries of the modeling domain, it has little impact on the results in the high SO_2 source regions. A nonlinear response of wet deposition to reductions in SO_x emissions is also predicted by RADM in emission reduction studies conducted using a preliminary version of the model.[29]

ADOM has also been used to investigate emission control scenarios for a winter episode in January and February of 1985. The results of this study, described in detail in an internal report, can be briefly summarized as follows.[24] A 50% reduction in SO_2 and SO_4 emissions across the modeling region shown in Figure 13 resulted in only a 25% reduction in sulfur wet deposition over most of the region. This strongly nonlinear response is apparently related to the very low H_2O_2 levels that the model predicted for the episode. A reduction in both SO_x and NO_x emissions by 50% increased the nonlinearity of the system—reductions in sulfur wet deposition ranged from 15 to 20% over most of the grid. A likely explanation for this result is that more OH is available for the gas phase oxidation of SO_2 when NO_x emissions are also reduced.[30]

These limited results show that most of the expected features of the acidic deposition system are well predicted by both RADM and ADOM. Data collected during the 1988 intensive field studies will be crucial in further validation of these models, which is a prerequisite to their use in exploring source-receptor relationships and determining optimal emission control strategies.

SUMMARY

Regional acid deposition modeling has progressed a great deal during the past 10 years, during which time several comprehensive models have been developed. These models incorporate the governing processes—gas and aqueous phase chemistry, transport, dry deposition, and cloud processes—in as much detail as possible. Two of these models, the Regional Acid Deposition Model (RADM), developed by the U.S. EPA, and the Acid Deposition and Oxidant Model (ADOM), funded by Environment Canada, Environment Ontario, the German Umweltbundesamt, and the Electric Power Research Institute, have undergone limited evaluation with observations. They have also been used to examine the response of regional sulfur deposition to projected reductions in SO_2 emissions. These models are currently being evaluated against data collected during field studies conducted in 1988. Preliminary results from this evaluation indicate that there are no major gaps in our understanding of the acid deposition system.

Comprehensive models, by their very nature, require large numerical resources. ADOM, for example, runs at about 5% of real time on a Cray-XMP. Thus, at this time, it is not practical to use these models to estimate the annual averages of deposition required to examine source-receptor relationships. For the time being, we must rely on simpler semi-empirical models to provide this information. Recall that most models applied to date can be classified as semi-empirical. These models emphasize the mechanistic relationships among the governing processes. However, the processes themselves are highly parameterized; in effect, little of our prior understanding of the processes is incorporated into the model. The required information on the processes is often contained in the *values* of the associated parameters, which are derived by fitting model predictions of wet deposition and ambient concentrations to observations.

Both comprehensive and semi-empirical models indicate that a large fraction of the sulfate observed in rain is formed by aqueous phase oxidation of SO_2 dissolved in cloud water. This also means that wet deposition of sulfur is limited by the availability of oxidants, the most important of which is believed to be hydrogen peroxide. Because sulfur wet deposition is oxidant limited, there is good reason to believe that changes in wet deposition will not be proportional to emission changes—the sulfur wet deposition system is nonlinear. Limited comprehensive modeling studies have also shown that the reduction in sulfur wet deposition is less than proportional to SO_2 emission reductions, particularly near source regions. These studies have also shown a more nonlinear response in winter than in spring of sulfur wet deposition to SO_2 emission reductions. Although more studies must be conducted to confirm a seasonal trend in nonlinearity, these results are consistent with our understanding of the sulfur wet deposition system—the reduced photochemical activity in winter results in extremely low levels of important aqueous phase SO_2 oxidants such as hydrogen peroxide.

While the nonlinearity in sulfur wet deposition must be considered in designing SO_2 emission control strategies, it is important to note that acidic deposition consists of both dry and wet deposition. Results from a semi-empirical model that accounts for oxidant limitation show that dry deposition responds almost proportionally (more than proportionally in some regions) to emission changes, while wet deposition shows less than a proportional response.[17] Because dry deposition is dominant in the source dominated areas where wet deposition has its largest values, total sulfur deposition is expected to respond more linearly than wet deposition to emission changes. Furthermore, the nonlinearity in sulfur wet deposition is important for space (within 100 to 200 km of source regions) and time (episodes) scales that are small relative to those considered in legislative proposals for emission reductions.

Some of the conclusions on source-receptor relationships drawn here are tentative because they are based on results from semi-empirical models. More definite conclusions will be provided by comprehensive models. However, for this to happen we must wait for advances in computing hardware as well as software to allow these models to run at speeds an order of magnitude higher than they do now.

REFERENCES

1. National Acid Precipitation Assessment Program, Washington, D.C. "Relationships Between Atmospheric Emissions and Deposition/Air Concentrations," SOS/T Report No. 8 (1990).
2. Eliassen, A. and J. Saltbones. "Decay and Transformation Rates of SO_2 as Estimated from Emission Data Trajectories and Measured Air Concentration," *Atmos. Environ.* 9:425–429 (1975).
3. Eliassen, A. and J. Saltbones. "Modeling of Long Range Transport of Sulphur Over Europe: A Two-Year Model Run and Some Model Experiments," *Atmos. Environ.* 17:1457–1473 (1983).
4. Eliassen, A. and O. Hov. "Status of Calculations of a European Country-by-Country NO_x Budget," EMEP MSC-W Note 1/87, the Norwegian Meteorological Institute, Oslo, Norway (1987).
5. Bhumralkar, C. M., W. B. Johnson, R. L. Mancuso, R. A. Thuillier, D. E. Wolf, and K. C. Nitz. "Interregional Exchanges of Airborne Sulfur Pollution

and Deposition in Eastern North America," in *Proceedings of the Second Joint Conference on Applications of Air Pollution Meteorology*, American Meteorological Society, New Orleans, LA (1980), pp. 225 to 231.

6. Samson, P. J. and M. J. Small. "Atmospheric Trajectory Models for Diagnosing the Sources of Acid Precipitation," in *Modeling of Total Acid Precipitation Impacts. Acid Precipitation Series*, Vol. 9 (Boston, MA: Butterworth Publishers, Inc., 1984), pp. 1-24.
7. Ellenton, G., B. Ley, and P. K. Misra. "Treating Exponential Mass Decay When Wet Scavenging Varies Discretely Within an Expanding Gaussian Dispersed Puff," in *The Meteorology of Acid Deposition, Perry J. Samson, Ed.* (Pittsburgh, PA: Air Pollution Control Association, 1983), pp. 528-536.
8. Ellenton, G., B. Ley, and P. K. Misra. "A Trajectory Puff Model of Sulfur Transport for Eastern North America," *Atmos. Environ.* 19:727-737 (1985).
9. Bolin, B., and C. Persson. "Regional Dispersion and Deposition of Atmospheric Pollutants with Particular Application to Sulfur Deposition Over Western Europe," *Tellus* 27:281-310 (1975).
10. Sheih, C. M. "Application of a Statistical Trajectory Model to the Simulation of Sulfur Pollution over Northeastern United States," *Atmos. Environ.* 11:173-178 (1977).
11. Shannon, J. "A Model of Regional Long-Term Average Sulfur Atmospheric Pollution, Surface Removal, and Net Horizontal Flux," *Atmos. Environ.* 15:689-701 (1981).
12. Fisher, B. E. A. "The Calculation of Long-Term Sulfur Deposition in Europe," *Atmos. Environ.* 12:489-501 (1978).
13. Rodhe, H. and J. Grandell. "On the Removal Time of Aerosol Particles from the Atmosphere by Precipitation Scavenging," *Tellus* 24:442-454 (1972).
14. Venkatram, A., B. E. Ley, and S. Y. Wong. "A Statistical Model to Estimate Long-Term Concentrations of Pollutants Associated with Long-Range Transport," *Atmos. Environ.* 16:249-257 (1982).
15. Smith, F. B. "The Significance of Wet and Dry Synoptic Regions on Long-Range Transport of Pollution and Its Deposition," *Atmos. Environ.* 15:863-873 (1981).
16. Venkatram, A. and J. Pleim. "Analysis of Observations Relevant to Long-Range Transport and Deposition of Pollutants," *Atmos. Environ.* 19:659-667 (1985).
17. Venkatram, A., K. Kashanian, P. K. Karamchandani, and T. Avitzur. "Probing the Acid Deposition System with a Semi-empirical Model: The Role of Oxidant Limitation," *Atmos. Environ.* 24A:125-131 (1990).
18. Shannon, J. D. and B. M. Lesht. "Estimation of Source-Receptor Matrices for Deposition of NO_x-N," *Water Air Soil Pollut.* 30:815-824 (1986).
19. Brodzinsky, R., B. K. Cantrell, R. M. Endlich, and C. M. Bhumrakkar. "A Long-Range Air Pollution Transport Model for Eastern North America. II. Nitrogen Oxides," *Atmos. Environ.* 18:2361-2366 (1984).
20. Chang, J. S., R. A. Brost, I. S. A. Isaksen, S. Madronich, P. Middleton, W. R. Stockwell, and C. J. Walcek. "A Three-Dimensional Eulerian Acid Deposition Model: Physical Concepts and Formulation," *J. Geophys. Res.* 92:14,681-14,700 (1987).
21. Venkatram, A., P. K. Karamchandani, and P. K. Misra. "Testing a Comprehensive Acid Deposition Model," *Atmos. Environ.* 22:737-747 (1988).
22. Easter, R. C., M. T. Dana, J. M. Thorp, K. M. Busness, J. M. Hales, G. S. Raynor, C. M. Benkovitz, R. L. Tanner, and J. D. Shannon. "Overview of the

Oxidation and Scavenging Characteristics of April Rains (OSCAR) Experiment," PNL-4869/UC-11, Pacific Northwest Laboratory, Richland, WA (1984).

23. Venkatram, A. and P. K. Karamchandani. "Source-Receptor Relationships. A Look at Acid Deposition Modeling," *Environ. Sci. Technol.* 20:1084–1091 (1986).
24. "Final Report: Winter Episode Study," ENSR 5210-004-100, ENSR Consulting and Engineering, Camarillo, CA (1989).
25. Walcek, C. J. and C. Berkowitz. "Modeling of Wet Scavenging and Subgrid-Scale Vertical Transport by Clouds in a Tropospheric Chemical Model," in *Air Pollution Modeling and Its Application VII*, H. van Dop, Ed. (New York: Plenum Publishing Corporation, 1989), pp. 267–275.
26. Middleton, P., J. S. Chang, J. C. del Corral, H. Geiss, and J. M. Rosinski. "Comparison of RADM and OSCAR Precipitation Chemistry Data," *Atmos. Environ.* 22:1195–1208 (1988).
27. Karamchandani, P. K., G. Kuntasal, and A. Venkatram. "Results from a Comprehensive Acid Deposition Model: Predictions of Regional Ozone Concentrations," in *Air Pollution Modeling and Its Application VII*, H. van Dop, Ed. (New York: Plenum Publishing Corporation, 1989), pp. 321–331.
28. Misra, P. K., R. Bloxam, C. Fung, and S. Wong. "Non-Linear Response of Wet Deposition to Emissions Reduction: a Model Study," *Atmos. Environ.* 23:671–687 (1989).
29. "Preliminary Evaluation Studies with the Regional Acid Deposition Model (RADM)," NCAR/TN-265 + STR, National Center for Atmospheric Research, Boulder, CO (1986).
30. Rodhe, H., P. Crutzen, and A. Vanderpol. "Formation of Sulfuric and Nitric Acid in the Atmosphere During Long-Range Transport," *Tellus* 33:132–141 (1981).

CHAPTER 8

A Modeling Investigation of Nonlinearities in the Wet Removal of SO_2 Emitted by Urban Sources

Jeremy M. Hales

INTRODUCTION

During the past several years, the possibility of nonlinearities in the atmospheric source-receptor sequence has emerged as an important issue in the determination of pollution emission-control policy. This importance is rooted in the expectation, or at least the suspicion, that decreases in pollutant emissions may not be accompanied by proportionate reductions in exposure at receptor sites. If such nonlinear behavior were indeed to occur, then cost-benefit analysis of an emission-reduction strategy would be, under the best of circumstances, complicated enormously. At worst, such behavior might lead to implementation of an ill advised strategy that, while resulting in considerable financial expenditure, would be largely ineffectual.

Obviously, there is a strong incentive to determine the extent and nature of nonlinearity for atmospheric source-receptor systems, and a number of theoretical and measurement studies have been conducted for this purpose. In particular, Easter and Luecken[1] have performed a modeling study wherein they varied storm inflow concentrations of SO_2 and SO_4^{2-} and computed corresponding wet deposition rates. The authors' simulated storms described nonreactive scavenging of sulfur oxides as well as the oxidation of aqueous S(IV) by H_2O_2 and O_3, and indicated moderate-to-extensive nonlinearity over rather wide ranges of conditions. The thrust of the present chapter is to provide corresponding calculations for a similar storm, which ingests SO_2 from a local urban source in addition to background pollutants, with the intent of examining nonlinear features of this more localized urban system.

Nonlinear atmospheric systems are usually rather complex in nature, and conceptual elements of this topic will not be elaborated here. For present purposes, it is sufficient to note that (under somewhat ideal circumstances) linear systems obey the relationship:

$$E_{\theta,i,j}(x,y) = A_\theta(x,y)\, S_{\theta,i,j} \qquad (1)$$

while nonlinear systems do not. Here $E_{\theta,i,j}$ represents the exposure resulting from some atmospheric process which has conveyed pollutant i from source j, to a receptor at position (x,y) over a time period (averaging time) θ. Examples of this quantity are surface concentration and pollutant wet-deposition amount. $S_{\theta,i,j}$ represents the corresponding pollutant input to the atmosphere such, for example, as a composite emission source or the chemical inflow to a storm from clear air. A_θ is simply a proportionality constant, which may vary with position, but not with $S_{\theta,i,j}$. Linear systems, by this criterion, obey Equation 1 at all points across the receptor domain; thus, the areal average of $E_{\theta,i,j}$ for a linear system obeys an equation of this general form as well. Figure 1 provides a graphical comparison of some linear and nonlinear systems.

Equation 1 suggests that the variable

$$L^* = \frac{S_{\theta,i,j} \partial E_{\theta,i,j}}{E_{\theta,i,j} \, \partial S_{\theta,i,j}} \approx \frac{(E_{\theta,i,j}|_2 - E_{\theta,i,j}|_1)/(E_{\theta,i,j}|_2 + E_{\theta,i,j}|_1)}{(S_{\theta,i,j}|_2 - S_{\theta,i,j}|_1)\ /(S_{\theta,i,j}|_2 + S_{\theta,i,j}|_1)} \quad (2)$$

should provide a useful quantitative measure of nonlinearity. For a totally linear system, the linearity parameter L* equals 1.0. L* exceeds 1.0 for systems that deviate from linearity in a concave-upward fashion (cf. curves ϵ and θ in Figure 1b). Systems that deviate from linearity in a concave-downward fashion (curves δ and η in Figure 1b) are characterized by L* less than 1.0. Equation 2 will be used as the primary indicator of the degree of nonlinearity in the modeling analysis presented below.

OVERVIEW OF MODELING APPROACH

The basic approach of the present analysis is to apply a two-dimensional reactive-storm model, which approximates the meteorological and chemical behavior of a warm-frontal storm, to an urban-pollution situation. The model storm continually ingests background pollution through its boundaries and processes this material by chemical reaction and precipitation scavenging. This storm is allowed to traverse a simulated SO_2 area-source, whose magnitude (53 mol/sec) and areal extent (55 km × 55 km) were chosen to be similar to those of the greater St. Louis, MO area. SO_2 from this source is allowed to mix, react, and wet-deposit with the previously noted background pollutants; both the urban SO_2 and its reaction product, SO_4^{2-}, have been numerically "tagged," however, to allow distinction between species of urban and background origin. The linearity parameter L* is derived from these simulations by performing multiple code executions for varying magnitudes of the urban source-strength, noting the associated wet deposition, and applying Equation 2.

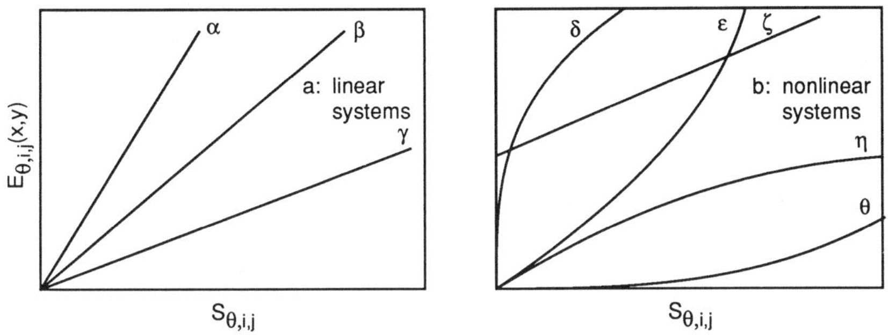

Figure 1. Examples of (a) linear and (b) nonlinear source-receptor behavior for a single receptor located at point (x,y).

OVERVIEW OF REACTIVE STORM MODEL

The reactive storm model used for this analysis is the PLUVIUS II code, which was applied in the previously noted Easter-Luecken study[1] and has been described by Hales.[2] Except for some minor modifications to the storm's flow field, the background simulation applied here is identical to the example in Hales' earlier paper. Initial storm chemistry was synthesized by allowing the code to execute, ingesting background pollution only, until an approximate steady-state condition was attained. Chemical species processed by this storm simulation are summarized in Table 1, and physicochemical mechanisms active in the model are noted in Table 2.

The wind field applied for these simulations is shown in Figure 2, while selected meteorological fields computed by the model are exhibited in Figures 3a to 3c. Without going into prolonged discussion, we note that these fields capture many of the basic features normally associated with warm-frontal storms, although numerous approximations and idealizations are apparent. Warm air from the left moves up and over the front, leading to water condensation, ice formation, and the onset of precipitation. Cold air approaches at low levels from the right, ascends in the vicinity of the front, and exits aloft. Figure 2 represents the wind field with respect to the front's horizontal position, which, in these simulations, is progressing to the right at 11 m/sec. Thus, the surface air at the right extremity of the plot is almost motionless with respect to the ground, while that on the left extremity is moving rapidly to the right.

Examples of chemical fields existing in the model immediately before introduction of the urban source are shown in Figures 3d to 3l. These plots correspond to "snapshots" of the chemical fields at one point in time and will modify somewhat at later times for two reasons. The first of these is that the modeled system contains several intrinsic nonlinear features, which lead to some degree of limit cycling in the extended solutions. The second reason for background field modification stems from its competitive interactions with urban pollutant during the storm's encounter with this additional pollution source.

Table 1. Chemical Species Included in Reactive Storm Model Version Used in Current Study

Background (Species Entering Model Domain From Left and Right Boundaries)
SO_2
SO_4^{2-} aerosol
HNO_3
NO_3^- aerosol
O_3
H_2O_2
Urban Source
Tagged urban SO_2
Reaction Products and Dissolved Species
SO_4^{2-} (from background aerosol and oxidized background SO_2)
S(IV) (dissolved background SO_2)
Tagged SO_4^{2-} (from oxidized urban SO_2)
Tagged S(IV) (dissolved urban SO_2)
NO_3^- (from background aerosol and HNO_3)
H_3O^+ (computed from semiempirical ion balance)
O_3
H_2O_2

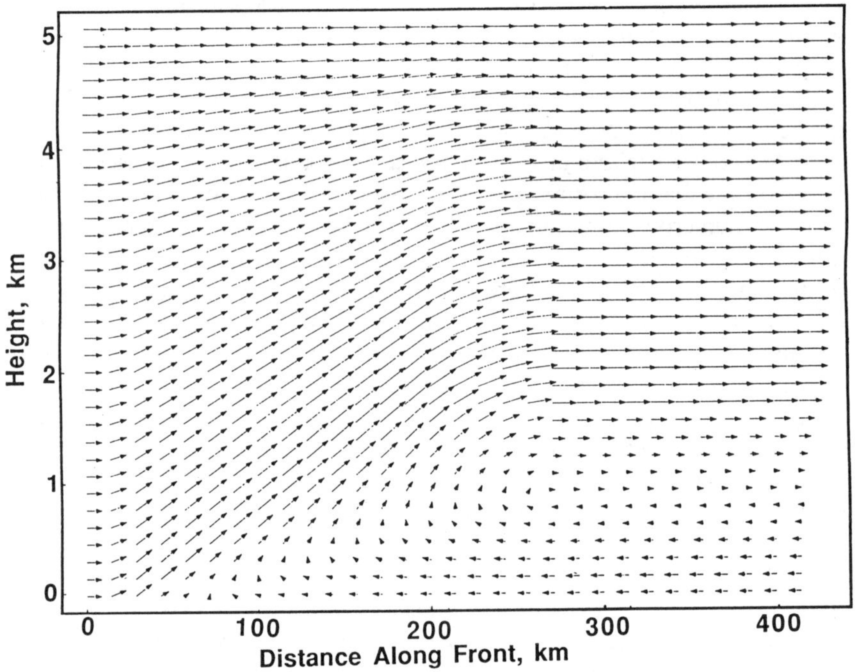

Figure 2. Wind field used for frontal-storm simulation.

Table 2. Active Mechanisms in Storm Model Version Used in Current Study

Cloud Physics		
	Cloud droplet autoconversion	
	Cloud droplet accretion	
	Cloud droplet riming	
	Raindrop freezing	
	Vapor deposition to ice	
	Raindrop evaporation	
Pollutant Uptake		
	Nucleation (both SO_4^{2-} and NO_3^- particles)	
	Aerosol capture by rain	
	Aerosol capture by snow	
	Gas dissolution in cloud and rain water:	
		SO_2
		HNO_3
		O_3
		H_2O_2
Reaction in Cloud and Rain Water		
	$S(IV) + OP_3 = SO_4^{2-}$	
	$S(IV) + H_2O_2 = SO_4^{2-}$	

DISTRIBUTION OF URBAN POLLUTANT

As was noted previously, urban SO_2 and SO_4^{2-} have been tagged numerically in this application of the code, so that their interactions within the storm can be followed in an unambiguous manner. Figure 4 presents a time series of storm cross sections, indicating the mixing-ratio fields of urban SO_2 as the storm traverses the city. Here the positions of the cloud and precipitation regions (cf. Figure 3) are indicated by the shaded regions, and the current position of the city is shown by the dark rectangle.

Figure 4 (top, left), corresponding to 2 hr after the first encounter of the urban source with the storm, shows a relatively confined SO_2 distribution. The urban SO_2 plume expands as it begins to be affected by the storm's more pronounced wind shears and vertical velocities. After 8 hr, the city is in the vicinity of the surface front and the plume is becoming extensively incorporated into the condensing cloud region. At 12 hr, the storm has completely traversed the city, and only residual pollution from the urban source remains.

Figure 5 presents similar plots for SO_4^{2-}, existing in cloud water and as aerosol, that has resulted from in-storm oxidation of urban SO_2. Here, however, only the 10- and 12-hr profiles are shown, owing to the low abundances of SO_4^{2-} at earlier times. Figure 6 shows corresponding 8-, 10-, and 12-hr plots for rainborne SO_4^{2-}. Comparing these figures with Figure 4 indicates that mixing of the urban SO_2 with the condensing cloud system is necessary for effective conversion of SO_2 to SO_4^{2-}, and its subsequent incorporation into rain. Direct incorporation of urban SO_2 into rain and its subsequent oxidation by H_2O_2 and O_3 seem to be of secondary importance.

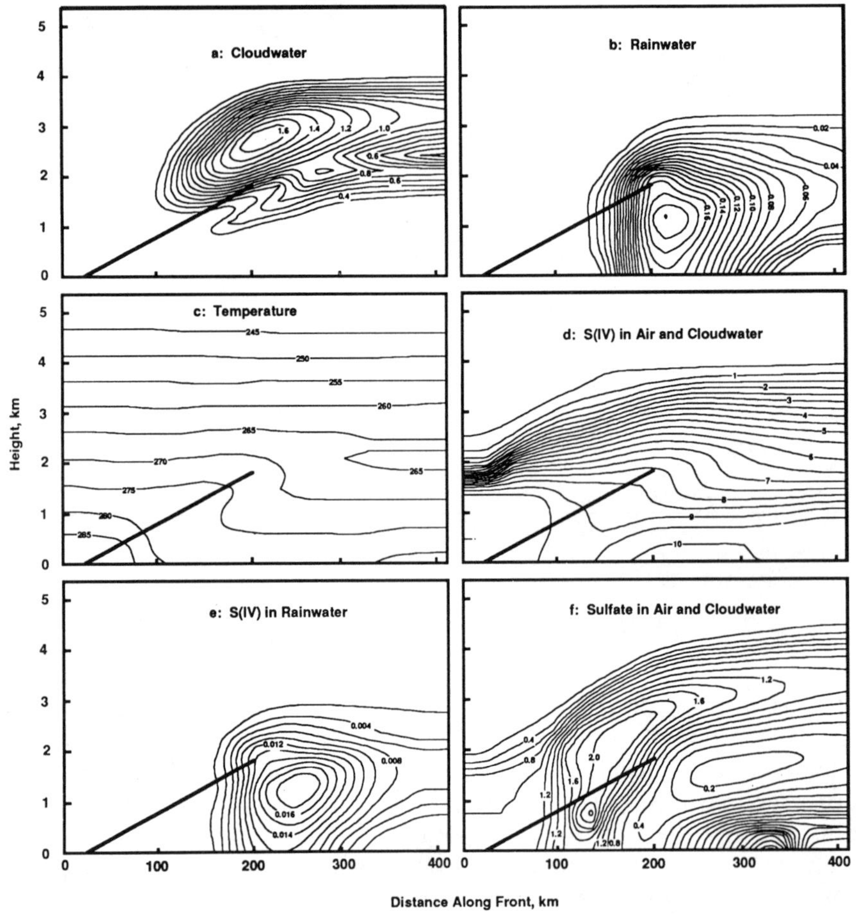

Figure 3. Selected meteorological and chemical fields from model output. Heavy diagonal line indicates front position. Moisture contours are molar mixing ratio × 10^3; chemical contours are molar mixing ratio × 10^9 (ppb); temperature contours are degrees K. Chemical fields correspond to the time just prior to introduction of the urban source. Meteorological fields are held constant throughout the computation. (a) Cloudwater; (b) rainwater; (c) temperature; (d) S(IV) in air and cloudwater; (e) S(IV) in rainwater; (f) sulfate in air and cloudwater; (g) sulfate in rainwater; (h) nitrate in air and cloudwater; (i) nitrate in rainwater; (j) hydrogen peroxide in air and cloudwater; (k) hydrogen ion in cloudwater; (l) hydrogen ion in rainwater.

POLLUTANT WET DEPOSITION

Total pollutant wet-deposition rates were computed from the primary model output by integrating surface deposition fluxes across the total storm. Results of these integrations, corresponding to calculations for the base source strength of 53 mol/sec, are given in Table 3, which shows deposition rates for background SO_4^{2-}, NO_3^-, and H_2O_2, as well as for urban SO_4^{2-}. In agreement with

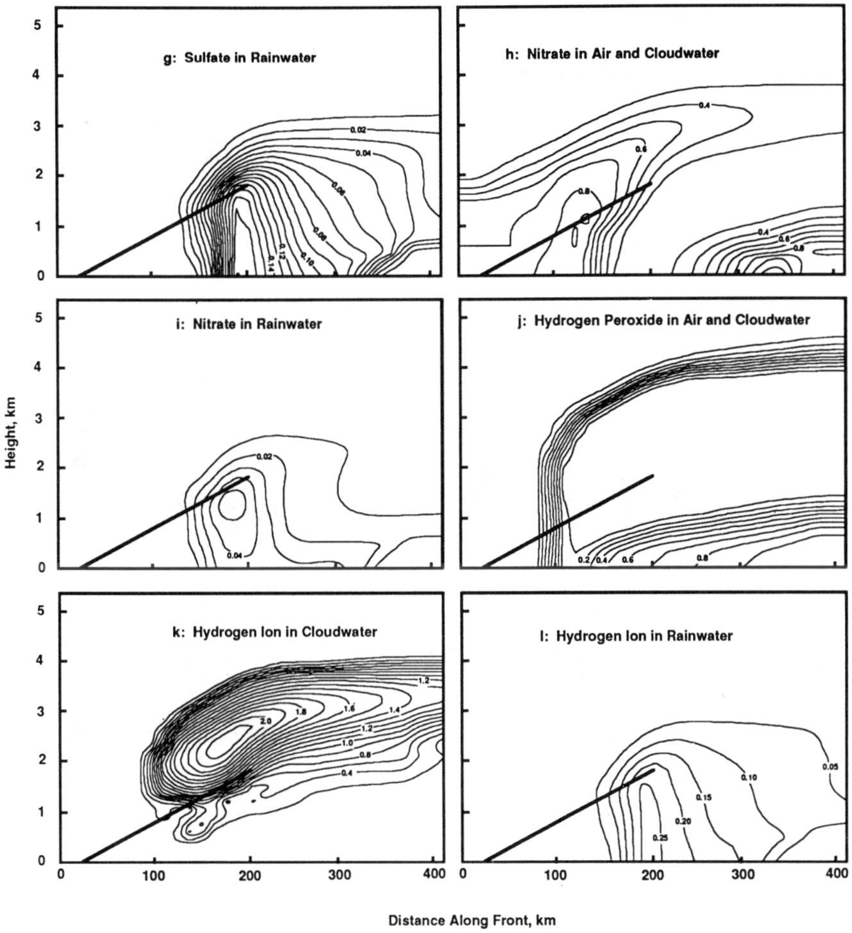

Figure 3 (continued).

the contour plots of Figure 6, Table 3 indicates that integrated wet-deposition of urban SO_4^{2-} increases rapidly during later stages of the storm interaction. At 12 hr, the urban contribution has risen to about 4% of the total SO_4^{2-} wet deposition.

Table 3 also gives some indication of the stability of the background pollution fields. Both SO_4^{2-} and NO_3^- appear to be increasing slightly, possibly as the result of residual transient effects from the initialization process. H_2O_2 seems to be affected to some extent by the presence of urban SO_2, exhibiting a marked decrease at about 6 hr into the encounter.

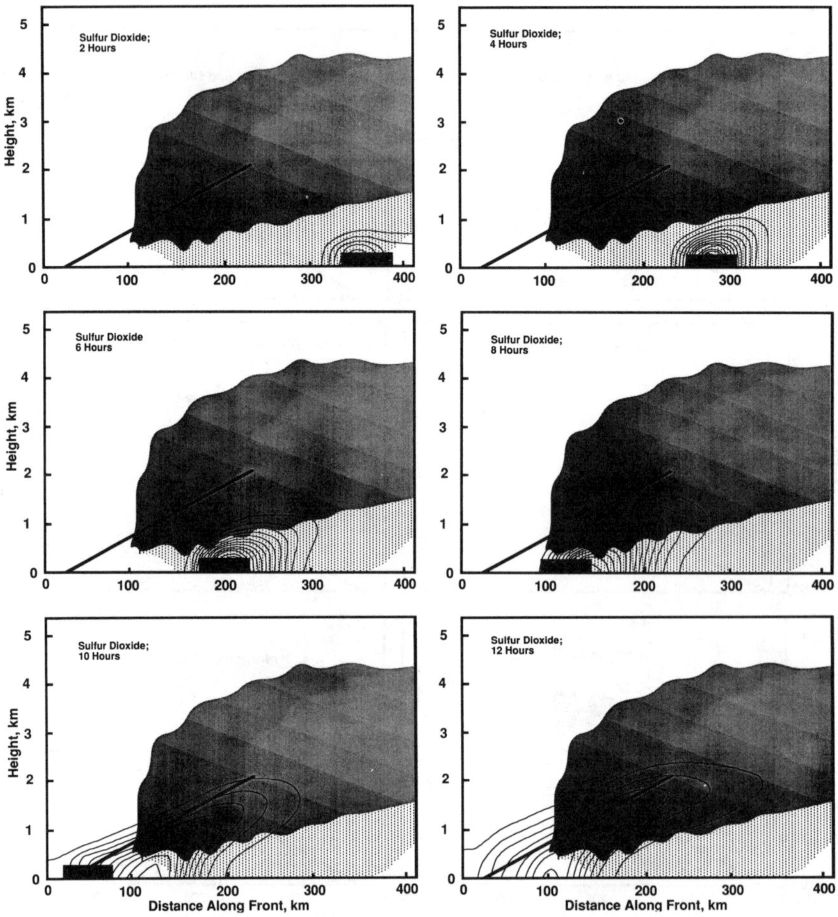

Figure 4. SO_2 mixing ratio contours resulting from an urban pollution source at progressive times after initial encounter. Dark rectangle indicates current position of urban source. Contour level = 0.5 ppb.

LINEARITY EVALUATION

The linearity parameter L* was determined by repeating the computations leading to Table 3 for a source strength slightly different than that for the base condition and inserting the integrated deposition rates and corresponding source strengths for $E_{\theta,i,j}$ and $S_{\theta,i,j}$, respectively, in Equation 2.

The results of this calculation, which are exhibited in Figure 7, were somewhat unexpected. As can be noted from this figure, pronounced nonlinearity is indicated during the initial storm encounter, but the system exhibits almost complete linearity during later stages. Since the preponderance of wet removal occurs during the period when L* ≈ 1.0, the total time-integrated exposure to

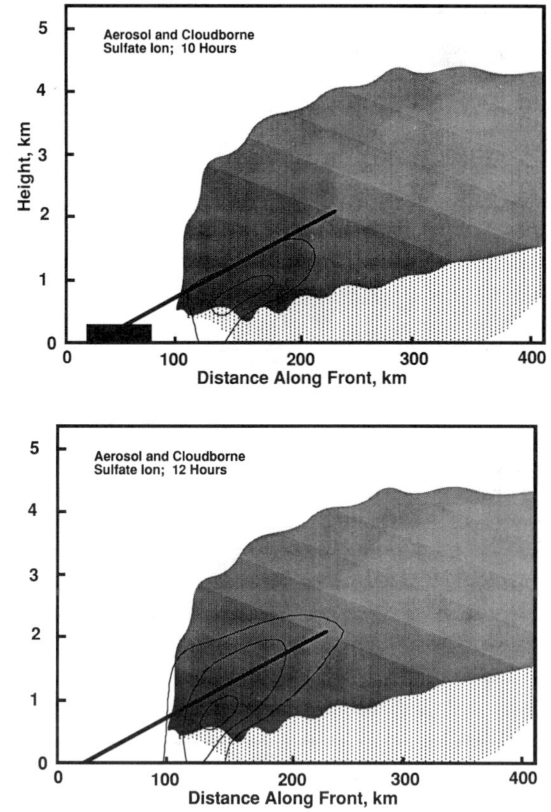

Figure 5. Aerosol and cloudborne SO_4^{2-} mixing ratio contours resulting from an urban pollution SO_2 source at progressive times after initial encounter. Dark rectangle indicates current position of urban source. Contour level = 0.05 ppb.

the storm must be considered as essentially linear as well—a result that initially seems rather paradoxical, in view of Easter and Luecken's contrary findings for the wet-removal of background pollution.

This apparent disagreement can be explained, however, by noting that the

Table 3. Storm-Integrated Deposition Rates Over Computation Domain (Moles per Second) as a Function of Time Since First Storm Encounter with Urban Source: Base-Case Urban SO_2 Emission Rate (53 moles/sec).

Time	Urban SO_4^{2-}	Background SO_4^{2-}	Background NO_3^-	Background H_2O_2
0	0	0.158E + 03	0.455E + 02	0.230E − 02
2	0.359E − 02	0.158E + 03	0.458E + 02	0.235E − 02
4	0.298E − 01	0.159E + 03	0.465E + 02	0.220E − 02
6	0.150E + 00	0.161E + 03	0.471E + 02	0.886E − 03
8	0.147E + 01	0.163E + 03	0.478E + 02	0.104E − 02
10	0.339E + 01	0.163E + 03	0.480E + 02	0.145E − 02
12	0.667E + 01	0.163E + 03	0.486E + 02	0.166E − 02

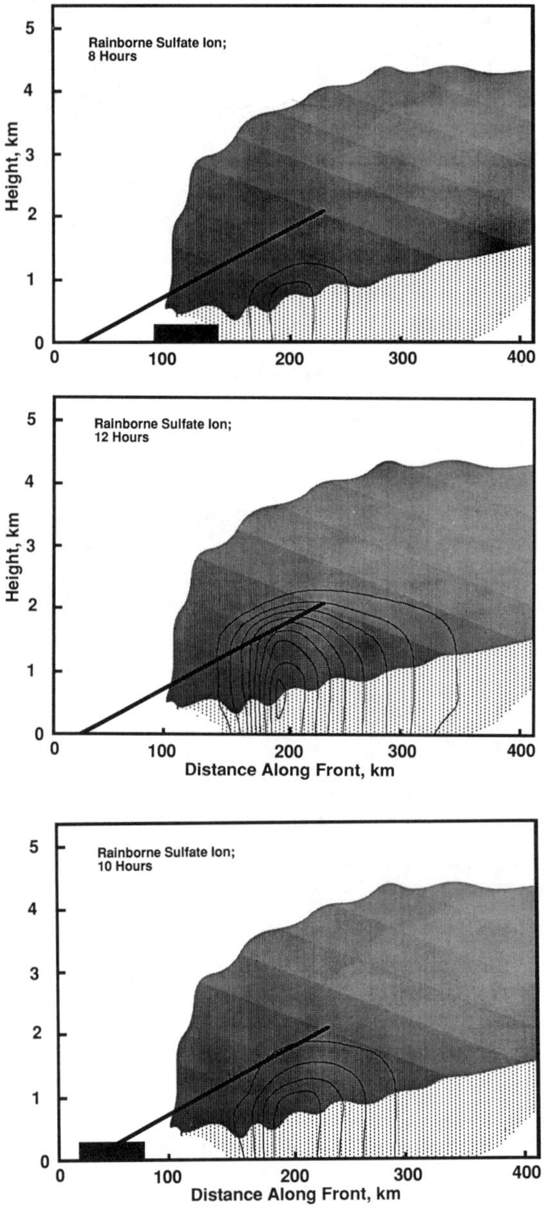

Figure 6. Rainborne SO_4^{2-} mixing ratio contours resulting from an urban pollution source at progressive times after initial encounter. Dark rectangle indicates current position of urban source. Contour level = 0.001 ppb.

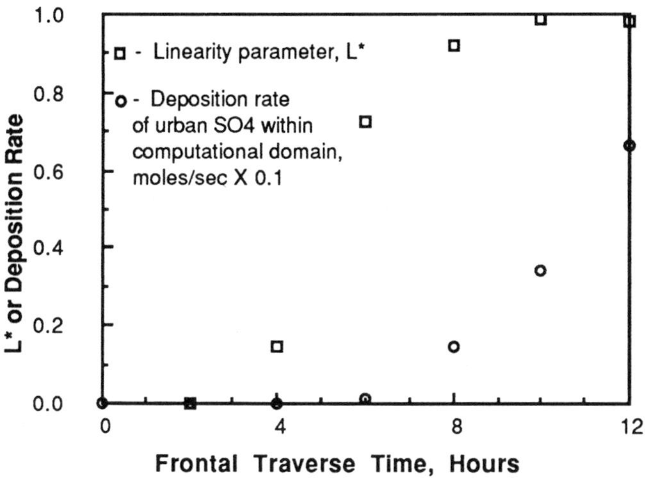

Figure 7. Total-storm wet-deposition rate for urban sulfate and associated linearity parameter, L*.

urban source contributes a small amount of SO_2 relative to the input of background material through the boundaries of the model domain. The major source of nonlinearity in this simulation involves reaction of aqueous S(IV) (dissolved SO_2) with H_2O_2. Within this environment a competition occurs for the oxidant pool between background and urban SO_2. From the urban SO_2 perspective this pool appears much larger and less limiting than it does to its larger background counterpart. Therefore, the urban calculations give a somewhat misleading indication of system linearity, even though the urban SO_2 interacts in the system in a manner that is chemically indistinguishable from the background burden.

This result has several rather important implications and can be examined in more detail using analytical calculations for a simple reactive box model, as follows. Consider a well-mixed flow reactor, where three reactive pollutants (denoted by the subscripts P, O, and T, indicating pollutant, oxidant, and tracer, respectively) are emitted continuously, with source-strengths S_P, S_O, and S_T. These pollutants are continuously removed by exit flows through the reactor boundaries and by the reactions

$$P + O \rightarrow P_{product}$$

$$T + O \rightarrow T_{product}$$

balancing the source terms to produce a steady-state situation.

Let the reaction rates of P, O, and T be described by the following equations:

$$\partial c_P/\partial t|_{reaction} = -k\ c_P\ c_O$$

$$\partial c_O/\partial t|_{reaction} = - k\ c_O\ (c_P + c_T)$$

$$\partial c_T/\partial t|_{reaction} = - k\ c_T\ c_O$$

The similarity between this simple zero-dimensional situation and the sulfur chemistry of the storm model described above is obvious. Giving the reactor a volume V and a mean residence time t*, one can perform material balances for the reactants and their products as follows:

$$S_P/V - c_P/t^* = k\ c_P\ c_O$$

$$S_O/V - c_O/t^* = k\ c_P + c_T)\ c_O$$

$$S_T/V - c_T/t^* = k\ c_T\ c_O$$

$$c_{P_{product}} = \frac{t^*\ S_P}{V} - c_P$$

$$c_{T_{product}} = \frac{t^*\ S_T}{V} - c_T$$

With some algebraic manipulation these forms can be rearranged to give explicit expressions for concentrations of the individual species and for the associated linearity parameters. Figure 8 provides results from a case example illustrating this system. Here source magnitudes have been chosen such that species P and O dominate, and T exists as a minor constituent acting as a tracer for P. As can be observed for the S-concentration curves as well as for the corresponding L* curves, the concentration of species $P_{product}$ is related to S_P in a moderately nonlinear manner, whereas the corresponding relationship for the tracer approximates linearity very closely. This is consistent with the earlier results for the more complex storm model and again indicates that great care must be taken in the design and interpretation of data from experiments that purport to measure system nonlinearity using tracer techniques.

CONCLUSIONS

The major finding of this study stems from a somewhat unexpected result. In circumstances when two chemically identical — but still distinguishable — species compete for a common reactant pool, any attempt to measure system response by varying the source strength of just one component will provide a misleadingly low indication of nonlinearity for the global system. This is not necessarily the case for systems whose nonlinearity derives from mechanisms other than competitions of this type. One can demonstrate easily, for example, that if nonlinearity in the modeled system were to stem solely from a simple second-order reaction, such as

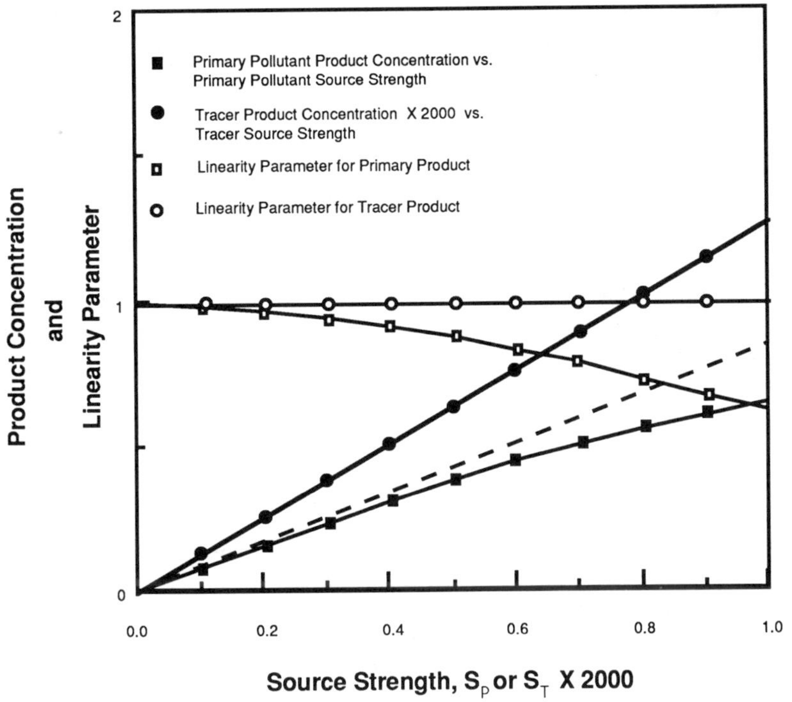

Figure 8. Example linearity parameters for a reactive species and its tracer in a well-mixed flow reactor. Results correspond to unit source strength of species O, unit reaction-rate constants, unit reactor volume, and a reactor residence time of 5.

$$\frac{dc}{dt} = - k c^2 \tag{3}$$

then observational or numerical tests such as the one described in this paper would provide valid indicators of total system nonlinearity. Unfortunately, second-order reactions of this type are encountered only infrequently in atmospheric chemistry, and in any event one is seldom sure of the exact mechanistic source(s) of nonlinearity in most practical situations. As a consequence, the approach illustrated in this paper must be considered a potentially misleading indicator of nonlinearity in a majority of circumstances.

This finding has rather profound implications for any physical or model-based test that proposes to gauge system nonlinearity by either: (1) adding a tracer to a known emission source (either a numerically tagged species or a physical tracer, such as $^{35}SO_2$) and varying the source strength of this labeled species or (2) numerically tagging a subset of a total emission field (say, Ohio's emissions within the field of total North American emissions) and varying this source strength.

In general, these types of tests should be expected to have the potential of yielding misleadingly low measures of total system nonlinearity, and their results should be carefully interpreted with this possibility in mind.

ACKNOWLEDGMENT

This work was supported by the U.S. Department of Energy (DOE) under Contract DE-AC06–76RLO 1830, at the Pacific Northwest Laboratory, which is operated for DOE by the Battelle Memorial Institute. Funding for this work was supplied via the DOE Processing of Emissions by Clouds and Precipitation (PRECP) program in conjunction with the National Acid Precipitation Assessment Program (NAPAP) Task Force. The author expresses his sincere appreciation to both the DOE and NAPAP for this support.

REFERENCES

1. Easter, R. C. and D. J. Luecken. "A Simulation of Sulfur Wet Deposition and its Dependence on the Inflow of Sulfur Species to Storms," *Atmos. Environ.* 22(12):2715–2739 (1988).
2. Hales, J. M. "A Generalized Multidimensional Model for Precipitation Scavenging and Atmospheric Chemistry," *Atmos. Environ.* 23(9):2017–2031 (1989).

CHAPTER 9

Chemical and Physical Processes in Clouds

Christian Seigneur

INTRODUCTION

The chemical and physical processes that occur in clouds and fogs affect significantly the formation of acidic species such as sulfate and nitrate. These sulfate and nitrate species may affect the environment in several ways.

If the cloud is a precipitating cloud, sulfate and nitrate species will be removed by precipitating cloud droplets or snow, thereby leading to acid precipitation. In mountainous areas, clouds may impact on trees and vegetation and wet deposition of acidity will occur. Fog droplets will to some extent settle and, then, lead to wet deposition.

If the cloud or fog evaporates, dissolved sulfate and nitrate species will return to an aerosol or gaseous phase. Sulfate and nitrate aerosols will affect atmospheric visibility by decreasing visual range and possibly causing some discoloration. Aerosols can also produce adverse health effects when inhaled and are consequently regulated under federal and state ambient air quality standards. Finally, acid aerosols and gases may deposit via dry processes thereby contributing to the total acid deposition budget.

It is therefore essential to understand the processes that govern the behavior of sulfate and nitrate species in clouds. In this chapter, the chemistry and physics of sulfate and nitrate species in clouds are reviewed. The mathematical models that simulate these processes are presented. Areas of uncertainties in these models are identified and further work is recommended to reduce these uncertainties. First, cloud chemistry is reviewed. Then, cloud physical processes are presented and their interactions with cloud chemistry are discussed.

CLOUD CHEMISTRY

Clouds affect the formation of sulfate and nitrate species as well as organic acids. The effect of clouds on ozone formation has been addressed elsewhere[1] and is not addressed here. We first address the cloud chemistry of sulfate, nitrate, and organic acid formation. Mathematical modeling of cloud chemistry is discussed next. Then, the effect of environmental conditions on sulfate and nitrate formation is reviewed. The effect of reductions in pollutant emissions on cloud chemistry concludes this section.

Sulfate Formation

The major chemical pathways leading to sulfate formation are summarized in Table 1.

The gas phase oxidation of SO_2 is dominated by the reaction of SO_2 with hydroxyl (OH) radicals. This reaction takes place primarily during the daytime when OH radical concentrations are high. The OH radicals are a result of atmospheric photochemical reactions and their concentrations will be highest when the solar flux and the concentrations of nitrogen oxides and reactive hydrocarbons are high.

The heterogeneous oxidation of SO_2 on carbon soot is not considered to be a major pathway for sulfate formation. Other reactions that lead to rapid oxidation of SO_2 to sulfate occur in the aqueous phase. The major aqueous phase oxidation routes for SO_2 include reaction with hydrogen peroxide (H_2O_2), ozone (O_3), oxygen (O_2), hydroxyl radicals (OH), and hydroperoxyl radicals (HO_2) and Cl_2^-. The oxygen reaction requires the presence of metal catalysts such as iron (Fe) and manganese (Mn). The aqueous oxidation of SO_2 by H_2O_2 generally predominates in the summertime when the H_2O_2 concentration reaches a few parts per billion (ppb). This reaction is very fast and its rate is not very sensitive to pH because the aqueous kinetics and S(IV) solubility vary in opposite directions as the pH varies. In the wintertime, H_2O_2 levels are lower (e.g., below 0.1 ppb) and this reaction is therefore less important. In the presence of dissolved iron, however, H_2O_2 may be formed and be a major contributor to sulfate formation, even in wintertime. The aqueous oxidation of SO_2 by O_3 takes place at a significant rate only in droplets with pH above about 4. This reaction is an important pathway for sulfate formation in raining clouds since the rainout of pollutants leads to higher pH values that favor this reaction. The reaction of SO_2 with O_2, catalyzed by trace metals, also occurs at a faster rate when the pH is high. Its kinetics is complex and synergism exists between the two trace metals. The rate of this reaction is a function of the concentrations of the trace metals. In raining clouds, trace metals will be removed by precipitation and this pathway will rapidly decrease in importance. The mechanisms of the reaction of SO_2 with OH, HO_2, and Cl_2^- are still uncertain and the importance of these pathways for sulfate formation depends strongly on the transfer of the OH and HO_2 radicals from the gas phase to the aqueous phase.

There are other pathways for the aqueous oxidation of SO_2 (e.g., nitrite ions, peroxyacetylnitrate, methyl hydroperoxide, peroxyacetic acid, and NO_2), however, these reactions are not considered to be major pathways for sulfate formation under most conditions.

Nitrate Formation

The major chemical pathways leading to nitrate formation are presented in Table 1.

The gas phase chemistry of NO_2 is governed by the OH radical chemistry during daytime and by the nitrate (NO_3) radical chemistry during nighttime. As mentioned above, OH concentrations are higher under conditions of pollution and high solar flux. The NO_3 radical is formed by reaction of NO_2 with O_3 and its concentration will therefore be higher under polluted conditions. During daytime, photolysis of NO_3 prevents any significant NO_3 levels. High concentrations of NO_3 occur therefore at night. In clouds, where the photochemical activity can be considerably reduced, the NO_3 concentrations may reach appreciable levels during the daytime as well.

The reaction of NO with OH leads to the direct formation of nitric acid (HNO_3). The nitrate radical can lead to nitric acid formation through several pathways. It can react with NO_2 to form nitrogen pentoxide (N_2O_5), which is subsequently hydrolyzed to HNO_3. This hydrolysis step takes place more efficiently as an heterogeneous reaction on the surface of water droplets or water-coated particulates. The NO_3 radical can also react with hydrocarbons such as aldehydes, aromatics, and olefins to form HNO_3 in the gas phase.

The aqueous phase chemistry of NO_2 is not significant for nitrate formation. However, dissolution of the nitrate radical in droplets may lead to its reduction by other chemical species (e.g., chloride, bisulfite, hydrogen peroxide) to form nitrate ions. This pathway may be important in clouds and fog when nitrate radical concentrations are important.

Organic Acid Formation

The two major organic acids found in clouds and fogs are formic acid and acetic acid. These are weak acids and, at equal concentrations, their contribution to the acidity of cloud/fog droplets is less than the contribution of sulfuric and nitric acids, which are strong acids. To date, the chemistry of organic acids has not been studied to the same extent as the chemistry of inorganic acids.

Formic acid is primarily formed by reaction of hydrated formaldehyde [$H_2C(OH)_2$] with OH radicals.[12,13] No significant formation of acetic acid was reported for wintertime simulations of polluted atmospheres.[13] Simulations of nocturnal stratus clouds in an urban atmosphere showed no significant formation of organic acids.[14]

Mathematical Models of Cloud Chemistry

Several mathematical models of the chemical kinetic mechanisms that govern sulfate and nitrate formation in clouds have been developed. Table 2 presents an overview of selected models. Such models range from simple models with only a few reactions to detailed chemical mechanisms with over

Table 1. Major Chemical Pathways of Atmospheric Sulfate and Nitrate Formation

Reaction	Rate Expression	Rate Constant[a] k_{298}	Rate Constant[a] $-E/R$	Ref.
Gas Phase				
$SO_2 + OH\ (+H_2O + O_2) \rightarrow H_2SO_4 + HO_2$	$k_o[SO_2][OH]$	1500 $ppm^{-1}min^{-1}$	—	2
$NO_2 + OH \rightarrow HNO_3$	$k_o[NO_2[OH]$	16300 $ppm^{-1}min^{-1}$	283	2
$NO_2 + O_3 \rightarrow NO_3 + O_2$	$k_o[NO_2][O_3]$	0.0474 $ppm^{-1}min^{-1}$	−1237	2
$NO_3 + RH^b \rightarrow HNO_3 + R$	$k_o[NO_3][RH]$	0.93 to 14,000 $ppm^{-1}\ min^{-1}$	—	2
$NO_3 + NO_2 \rightarrow N_2O_5$	$k_o[NO_2][NO_3]$	1780 $ppm^{-1}min^{-1}$	−30	2
$N_2O_5 \rightarrow NO_2 + NO_3$	$k_o[N_2O_5]$	3.12 min^{-1}	−5420	2
$N_2O_5 + H_2O \rightarrow 2HNO_3$	$k_o\ A/V\ [N_2O_5]$	2.07 min^{-1}	—	3
Aqueous Phase				
$S(IV) + H_2O_2 \rightarrow S(VI) + H_2O$	$\frac{k_o[H_2O_2][SO_2]}{1 + 16[H^+]}$	$7.6 \times 10^7\ M\ min^{-1}$	−4430	4
$S(IV) + O_3 \rightarrow S(VI) + O_2$	$(k_o[SO_2] + k_1[HSO_3^-] + k_2[SO_3^{2-}])[O_3]$	$1.44 \times 10^6\ M\ min^{-1}$ $2.22 \times 10^7\ M\ min^{-1}$ $9.00 \times 10^{10}\ M\ min^{-1}$	— −5530 −5280	5
$S(IV)(+\frac{1}{2}O_2) \xrightarrow{Mn^{2+}} S(VI)$	$S(IV) > 10^{-4}\ M$ $4 \times 10^4 \cdot 10^{-4.07\sqrt{I}/(1+\sqrt{I})}\ [Mn^{2+}]^2$ $S(IV) < 10^{-4}M$ $6 \times 10^4 \cdot 10^{-4.07\sqrt{I}/(1+\sqrt{I})}\ [Mn^{2+}][S(VI)]$		—	6
$S(IV)(+\frac{1}{2}O_2) \xrightarrow{Fe^{3+}} S(VI)$	$\frac{360\ 10^{-2\sqrt{I}/(1+\sqrt{I})}}{1 + 150[S(IV)]^{2/3}}\ \frac{[Fe^{3+}][S(IV)]}{[H^+]}$		—	7

$HSO_3^- + OH(+ HSO_3^- + O_2) \rightarrow SO_4^{2-} + SO_4^- + H_2O + H^+$	$k_o[OH][HSO_3^-] + k_1[OH][SO_3^{2-}]$	2.7×10^{11} *M* min^{-1} 3.1×10^{11} *M* min^{-1}	—	8
$SO_4^- + H_2O_2 \rightarrow HSO_4^- + HO_2$	$k_o[SO_4^-][H_2O_2]$	7.2×10^{8} *M* min^{-1}	—	9
$S(IV) + HO_2 \rightarrow S(VI) + OH$	$k_o[HO_2][HSO_3^-] + k_1[OH][SO_3^{2-}]$	10^{6} *M* min^{-1} 10^{5} *M* min^{-1}	—	10
$HSO_3^- + Cl_2^-(+ HSO_3^- + O_2) \rightarrow SO_4^{2-} + SO_4^- + 2H^+ + 2Cl^-$	$k_o[Cl_2^-][HSO_3^-]$	7×10^{7} *M* min^{-1}	—	8
$NO_3 + Cl^- \rightarrow NO_3^- + Cl$	$k_o[NO_3][Cl^-]$	6×10^{9} *M* min^{-1}	—	11
$NO_3 + HSO_3^-(+ HSO_3^- + O_2) \rightarrow NO_3^- + 2H^+ + SO_4^- + SO_4^{2-}$	$k_o[NO_3][HSO_3^-]$	6×10^{9} *M* min^{-1}	—	11
$NO_3 + H_2O_2 \rightarrow NO_3^- + H^+ + HO_2$	$k_o[NO_3][H_2O_2]$	6×10^{7} *M* min^{-1}	—	11

[a] $k = k_{298} \exp[-E/R(1/T - 1/298)]$, where k_{298} is reaction rate at 298 K and T is temperature in kelvin. For each reaction, the values of k_o, k_1, k_2, etc. are presented in the same order.

[b] RH represents reactive hydrocarbons such as aldehydes, aromatics, or olefins.

Table 2. Selected Mathematical Models of Cloud Chemistry[a]

Gas-Phase Chemistry[b]	Aqueous-Phase Chemistry[b]	Gas-Phase/ Aqueous-Phase Mass Transfer	Ref.
276 reactions	188 reactions	24 equilibria with diffusion-limited mass transfer	13
87 reactions	64 reactions	39 equilibria	10
71 reactions	43 reactions	18 equilibria, diffusion-limited mass transfer for radicals	15
22 reactions	40 reactions	13 equilibria with diffusion-limited mass transfer	11
None	15 reactions	10 equilibria	16
None	10 reactions	6 equilibria	17

[a]This list of mathematical models of cloud chemistry is not exhaustive, but represents the major types of models.
[b]Reactions include kinetic reactions, ionic equilibria, hydration equilibria, and precipitation equilibria.

100 reactions. Some models treat both gas phase and aqueous phase chemistry while others only treat aqueous phase chemistry. The latter models are inappropriate because gaseous species affect the aqueous phase chemistry through gas phase/droplet transfer. Clouds with a liquid water content of 1 g/m^3 consist of 99.9% air and 0.1% water. Although the partitioning of chemical species between the gas phase and the aqueous phase will depend on the species solubility (e.g., SO_2 is primarily in the gas phase, whereas HNO_3 is primarily in the aqueous phase), the gas phase species have a nonnegligible effect on the chemical composition of the aqueous phase. Table 3 presents an example of chemical species partitioning for a cloud with a liquid water content of 1 g/m^3 and droplet pH of 4.

Mathematical models that include a large number of chemical reactions[13] are extremely useful to analyze the chemical kinetics of sulfate and nitrate formation in detail. However, such models are difficult to integrate in acid deposition models that must also treat cloud dynamics, atmospheric transport, and dispersion and removal processes over large temporal and spatial scales. Mathematical models have therefore been developed to offer a compromise between computational requirements and mechanistic details.[10,15] Such models can be incorporated into large acid deposition models while still providing a reasonably accurate representation of cloud chemistry.

Effect of Environmental Conditions on Sulfate and Nitrate Formation

The earlier discussion of the chemical processes that affect the chemistry of sulfate and nitrate formation suggests that the environmental conditions (i.e.,

Table 3. Gas-Phase/Aqueous-Phase Partitioning of Major Chemical Species

	Fractional Partitioning[a]	
Chemical Species	**Gas Phase**	**Aqueous Phase**
SO_2	0.996	0.004
NO[b]	0.99999995	0.00000005
NO_2[b]	0.9999998	0.0000002
O_3[b]	0.9999997	0.0000003
H_2O_2[b]	0.37	0.63
OH	0.9994	0.0006
NH_3	0.07	0.93
H_2SO_4	0	1.0
HNO_3	0.00003	0.99997

Note: Liquid water content = 1 g/m^3; pH = 4; T = 298 K.

[a]Equilibrium between the two phases is assumed.
[b]Independent of pH.

meteorology, background pollution) are critical to the rate of sulfate and nitrate formation. Discussed in this section are the major factors that govern sulfate and nitrate formation. Major factors that are discussed here include season, daytime vs. nighttime, atmospheric liquid water content, and background pollution levels.

Season

The effect of season corresponds primarily to the effect of the solar flux on background pollution levels. Oxidants such as O_3, H_2O_2, and OH radicals reach higher concentrations in summer than in winter. For example, H_2O_2 levels will be of the order of a few parts per billion in the summertime, but will be less than 0.1 ppb in the wintertime. O_3 and OH radical concentrations show less dramatic variations than H_2O_2, but typical concentrations are significantly lower in wintertime than in summertime. Therefore, seasonal variation of sulfate and nitrate concentrations should be expected. This seasonal variation has been observed for sulfate concentrations. However, such seasonal variation is not as apparent for nitrate concentrations.[18,19]

Daytime vs. Nighttime

As discussed in the review of the chemical processes, atmospheric photochemistry plays a pivotal role in the formation of sulfate and nitrate. Formation of sulfate should be more important during daytime when OH radical concentrations and other oxidants are high. However, for cases where aqueous pathways dominate, oxidation of SO_2 to sulfate will not be as sensitive to the solar flux.

The nitrate chemistry is governed by OH chemistry during the daytime and by NO_3 chemistry during the nighttime. Both daytime and nighttime chemical mechanisms can lead to NO_2 oxidation rates of the order of 10%/hr.

Liquid Water Content

Under unsaturated conditions (i.e., with relative humidity below 100%), sulfate and nitrate formation are likely to be dominated by gas phase chemistry. It should be noted, however, that particulate chemistry can lead to significant rates of sulfate formation under conditions of high relative humidities. Nocturnal nitrate chemistry will also be affected by the presence of liquid-coated particulates because the rate of N_2O_5 hydrolysis is a function of the surface area of these particulates.

In saturated atmospheres (i.e., clouds, fog), aqueous chemistry will dominate sulfate formation. The gas phase oxidation of SO_2 leads to rates of sulfate formation of the order of 1%/hr; the aqueous oxidation of SO_2 by H_2O_2 is very rapid with a half-life for SO_2 which can be of the order of a few minutes. The oxidation of SO_2 by O_3 or O_2 can occur at a rate of the order of 10%/hr.

Nitrate formation in saturated atmospheres may be governed by NO_3 chemistry only if light intensity is low (i.e., in dark clouds or at night). In other cases, nitrate formation through gas phase chemistry of NO_2 and OH dominates nitrate formation.

Pollution Levels

Oxidant levels such as OH radicals, O_3 and H_2O_2 are functions of the concentration of photochemical smog precursors, i.e., nitrogen oxides (NO_x) and reactive hydrocarbons. Therefore, pollution levels are important in determining the rates of oxidation of SO_2 to sulfate and NO_2 to nitrate. Relative concentrations of SO_2 and NO_x will also affect the rate of sulfate formation. In a plume with high NO and NO_2 concentrations, OH radical concentrations are depleted and gas phase oxidation of SO_2 will be limited by the availability of OH radicals. By comparison, sulfate formation will occur faster in a plume that contains only SO_2 and no NO_x. For example, SO_2 oxidation occurs faster in smelter plumes that contain small amounts of NO_x than in power plant plumes that contain large amounts of NO_x.

Summary of the Effect of Environmental Conditions

Computer simulations of sulfate and nitrate formation have been performed by several investigators to identify the chemical pathways leading to sulfate and nitrate formation under a variety of environmental conditions. The results of these studies are summarized in Tables 4 and 5 for sulfate and nitrate, respectively.

Effect of Pollutant Emission Changes on Cloud Chemistry

Because sulfate and nitrate species have adverse effects on the environment through acid deposition, visibility degradation, and aerosol formation, it is of

Table 4. Summary of Major Chemical Pathways for Sulfate Formation

Environmental Conditions	Major Chemical Pathways	Ref.
Cloud-free atmosphere, daytime	$OH(g)$, $O_2(aq)$	20
Cloud-free atmosphere, nighttime	$O_2(aq)$, $H_2O_2(aq)$	20
Raining clouds, polluted continental background, summer, daytime	$H_2O_2(aq)$, $O_3(aq)$	21
Nonraining clouds, polluted continental background, summer, daytime	$OH(g)$, $H_2O_2(aq)$	21
Nonraining stratus clouds, urban area, spring, daytime	$H_2O_2(aq)$, $O_2(aq)$	15
Nonraining stratus clouds, urban area, spring, nighttime	$H_2O_2(aq)$, $O_2(aq)$, $OH(aq)$	15
Nonraining stratus clouds, industrial area, winter, daytime	$OH(g)$, $OH(aq)$	13
Nonraining stratus clouds, industrial area, winter, nighttime	$H_2O_2(aq)$, $O_2(aq)$	13
Plume in stratus clouds, spring, nighttime	$O_2(aq)$, $H_2O_2(aq)$, $OH(g)$	15
Fog, urban area, nighttime	$H_2O_2(aq)$, $O_2(aq)$	22
Nonraining cumulus clouds, polluted continental background, summer, daytime	$H_2O_2(aq)$, $HO_2(aq)$, $OH(aq)$, $Cl_2^-(aq)$	10
Nonraining stratus clouds, tropical marine background, summer, daytime	$H_2O_2(aq)$, $O_3(aq)$, $OH(aq)$	12
Nonraining stratus clouds, marine background, summer, daytime	$H_2O_2(aq)$, $O_3(aq)$, $OH(aq)$	11

particular interest to assess the relationships between the formation of sulfate and nitrate species and the levels of their precursors. Once such relationships are established, it is possible to develop effective pollutant emission control strategies. To that end, mathematical models have been applied to calculate the effect of precursor emission levels on sulfate and nitrate formation.

Seigneur et al.[25] first demonstrated the nonlinearities of the SO_2/sulfate relationship by simulating cloud chemistry with a fairly comprehensive chemical kinetic mechanism. (A nonlinear SO_2/sulfate relationship means that a reduction in SO_2 levels will not lead to a proportional reduction in sulfate formation.) The NO_2/nitrate relationship appeared to be nearly linear. These theoretical calculations were later extended to include actual data on back-

Table 5. Summary of Major Chemical Pathways for Nitrate Formation

Environmental Conditions	Major Chemical Pathways	Ref.
Cloud-free atmosphere, daytime	OH(g)	23
Cloud-free atmosphere, nighttime	NO_3(g), N_2O_5(aq)	23
Nonraining, stratus clouds, urban area, spring, daytime	OH(g), NO_3(aq), N_2O_5(aq)	15
Nonraining stratus clouds, urban area, spring, nighttime	NO_3(aq), N_2O_5(aq)	15
Nonraining stratus clouds, industrial area, winter, daytime	OH(g)	13
Plume in stratus cloud, spring, nighttime	OH(g), N_2O_5(aq), NO_3(g), NO_3(aq)	24
Fog, urban area, nighttime	OH(g)	22

ground pollution levels and meterological conditions over the northeastern U.S. The results confirmed the nonlinearity of the SO_2/sulfate relationship and its extent was calculated as a function of region and season.[26]

CLOUD PHYSICS

A variety of physical processes governs the formation of clouds, their evolution, precipitation, and evaporation. For example, the formation of cloud droplets, ice crystals, and rain drops, the transport of air in and out of clouds, and the scattering of light in and around clouds are physical processes that affect cloud characteristics. All these processes have a direct effect on the chemistry of clouds. This section does not present a detailed description of the physics of clouds, but rather discusses the physical processes of clouds as they affect cloud chemistry. We address successively atmospheric transport, photolytic processes, droplet size distribution, scavenging processes, and evaporation processes.

Transport Processes

Transport processes are a major component of the phenomenon of acid formation in clouds. The formation of sulfate and nitrate species requires that their respective precursors, SO_2 and NO_2, be brought in contact with oxidants in the cloud. The type of chemical species brought into the cloud will affect the pH of cloud droplets which in turn will affect the kinetics of acid formation. For example, if a significant amount of ammonia is transported into a cloud or fog, the droplet pH will rise and some reactions will occur more rapidly (e.g., oxidation of SO_2 by O_3 and O_2) while others may slow down. Also, transport processes will determine how long the chemical species reside in the cloud. In cases where the reaction of SO_2 or NO_2 oxidation is not very fast, the residence time of precursors will determine the amount of acid formed.

Another aspect of the importance of transport processes in clouds is the treatment of precipitation. Clearly, it is essential to be able to simulate precipitation accurately, since it will determine (1) the removal rate of pollutants from the cloud and, consequently, (2) the evolution of the chemical composition of the cloud as precipitation takes place.

Mathematical modeling of the dynamics of clouds is a complex problem. Stratiform clouds are generally more easily modeled than cumulus clouds because the former have a structured behavior compared to the convective behavior that characterizes the latter. The treatment of cumulus clouds has, therefore, been more challenging. Walcek and Taylor[16] proposed a model where air was entrained from the top of the cloud in addition to the typical updraft and entrainment of ambient air. Recently, Venkatram et al.[27] completed a model of cloud dynamics that represents the salient features of the physical processes involved.

The importance of transport processes in determining the chemical composition of clouds was demonstrated in the simulation of Los Angeles stratus clouds.[15] In this simulation, it appeared that the chemical composition of nocturnal stratus clouds was determined primarily by the chemical composition of the air entrained into the cloud and only to a limited extent by the chemical reactions taking place in the cloud. These results emphasize the need to understand and correctly treat the physical processes that govern the transport of pollutants into clouds.

Photolytic Processes

Photochemical reactions are the driving force of oxidant formation. Therefore, the treatment of photolytic processes may be critical to cloud chemistry for time scales where significant oxidant formation may occur. Inside the cloud, light intensity is typically lower than in clear air because of light extinction due to cloud droplets. At the edges of the cloud, however, light intensity may significantly increase due to light scattering by cloud droplets.[28] Therefore, the formation of hydrogen peroxide can be significantly enhanced at the edge of clouds where the photolytic flux is very high because high concentrations of HO_2 radicals, the precursors of H_2O_2, will result from active photochemistry.

The treatment of light intensity in clouds is a difficult problem. To date, most cloud models include a simple treatment of light intensity, such as an arbitrary factor to adjust light intensity in the cloud. This factor is assumed to be constant throughout the cloud. Clearly, such a simple treatment does not provide an accurate simulation of photochemical processes and may therefore affect the overall cloud chemistry. A stratus cloud model that includes a comprehensive treatment of atmospheric optics was developed recently by Jacob et al.[13] This model, however, is one-dimensional. Treatment of three-dimensional light scattering is significantly more complex. Models of plume visibility have addressed the problem,[29] however, such models apply only to optically thin atmospheres and, consequently, not to clouds or fogs. Models for simplified clouds, e.g., cubic clouds, exist, however, they provide a simplistic representation of the problem. Clearly, more work is needed in this area.

Droplet Size Distribution

Clouds and fog consist of droplets (and ice crystals at below freezing temperatures) that vary in size from condensation nuclei to large raindrops. These droplets constantly interact through a variety of physical processes (e.g., coagulation, accretion, riming, water vapor condensation, evaporation, condensation) which are schematically summarized in Figure 1. These droplets are also entrained through the clouds with, typically, cloud droplets moving upward with the ascending air and large rain drops moving downward. Thus, droplets will have different histories according to the microphysical and dynamic processes that govern their formation and transport, respectively.

Therefore, one may expect that the chemical composition of droplets will vary according to their history. This is clearly a complex problem due to the large number of chemicals and droplets involved. Typically, mathematical models of cloud chemistry assumed that the chemical composition of droplets is independent of their size and is only a function of time and location. Some models have used some broad distribution of droplets in size categories. For example, Tremblay and Leighton[17] treated cloud droplets and raindrops with different chemical composition. It would be of interest to assess the effect that the assumption of homogeneous droplet chemical composition has on the results of model simulations.

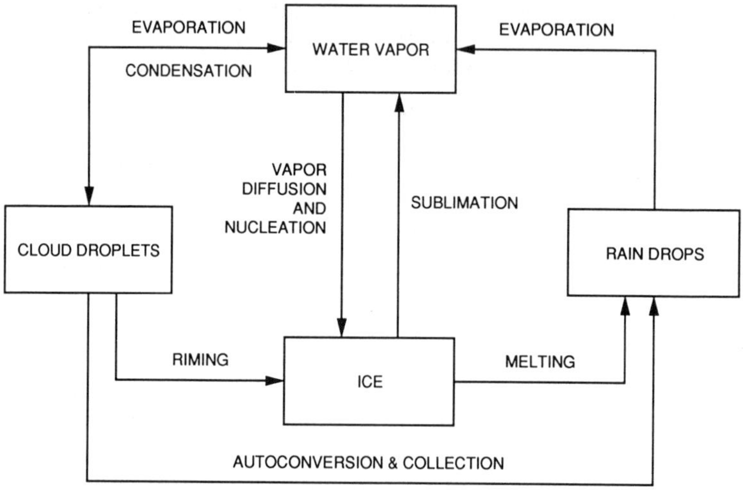

Figure 1. Schematic description of cloud microphysics.

Scavenging of Gases and Aerosols by Droplets

Clouds and fog form initially by condensation of water vapor on condensation nuclei. Gases and aerosols present in the interstitial air of clouds and fog will be transferred to the aqueous phase by several processes. Gases will diffuse from the bulk gas phase to the surface of droplets where partitioning between the gas and the aqueous phase will depend on the gas solubility. Aerosols will be transferred to the aqueous phase either (1) by acting as condensation nuclei or (2) by coagulating with an existing droplet.

Mathematical models of the transfer of gases to droplets have been developed. Schwartz[30] showed that for most gases and conditions, mass transfer was not rate limiting and equilibrium based on Henry's law could be used to calculate the partitioning between the gas phase and the aqueous phase. The treatment of mass transfer in cloud chemistry models is summarized in Table 2. The transfer of aerosols to droplets is more complex because coagulation processes depend on the size of both the aerosols and the droplets and water condensation processes depend on the chemical composition of the aerosols. Mathematical techniques exist to treat coagulation and condensation processes,[31] however, these techniques have not been included in cloud chemistry models. Most models either assume total dissolution of aerosols in droplets or use an arbitrary scavenging factor.

The scavenging of gases and aerosols by precipitating raindrops below clouds is also of particular interest. Most models of cloud chemistry and precipitation use simple treatments of the scavenging processes assuming a constant scavenging efficiency that can range from no scavenging to total scavenging. Experimental data suggest that aerosols in the small and large size range are preferentially scavenged and that aerosols in the 0.1 to 1 μm size range have the lowest scavenging efficiency. These results are in agreement with coagulation theory and models could therefore be developed to include a more detailed treatment of aerosols washout by precipitation than the existing treatments.

Evaporation of Clouds and Fogs

Most clouds and fogs do not precipitate, but instead evaporate leaving a myriad of gases and aerosols in the ambient unsaturated air. Chemical species will be present in gaseous or aerosol form according to their thermodynamic properties. Species such as sulfuric acid and seasalt will preferentially be present as aerosols, whereas species such as NO and NO_2 will be in the gas phase. HNO_3, ammonia, and SO_2 will be partitioned between the gas phase and the aerosol phase according to thermodynamics and environmental conditions. Mathematical models have been developed that describe aerosol chemical composition.[32,33] These models can also be used to calculate the aerosol size distribution that would result from the evaporation of a cloud or fog if one knows the cloud droplet chemical composition and size distribution.

CONCLUSION

The chemical and physical processes that govern the chemical composition of clouds were reviewed. Mathematical models of cloud chemistry range from simple to very detailed chemical mechanisms. Models that offer a compromise between accuracy and computational requirements are the most useful for integration into air quality models that must treat transport, dispersion, and deposition processes in addition to chemistry. Some uncertainties remain in kinetic mechanisms of cloud chemistry, particularly the aqueous phase chemistry and gas phase/aqueous phase mass transfer of free radicals.

Physical processes such as transport, photolysis, droplet scavenging of gases and aerosols, and evaporation of droplets have important effects on the chemistry of clouds. Because of their complexity, these processes are generally treated by means of simple models. It is necessary to develop further the treatment of physical processes in clouds and their coupling with chemical processes.

It is essential that theoretical representations of cloud processes be evaluated against experimental data. Such cloud model evaluation studies have been limited to date and additional work is needed in this area.

ACKNOWLEDGMENTS

The author is grateful to Pradeep Saxena, Electric Power Research Institute, for a few pertinent comments and several astute suggestions.

REFERENCES

1. Seigneur, C., and P. Saxena. "The Impact of Cloud Chemistry on Photochemical Oxidant Formation," *Water Air Soil Pollut.* 24:419–429 (1985).
2. Atkinson, R., and A. C. Lloyd. "Evaluation of Kinetic and Mechanistic Data for Modeling of Photochemical Smog," *J. Phys. Chem. Ref. Data* 13:315–444 (1984).
3. Harker, A. B., and D. R. Strauss. "Kinetics of the Heterogeneous Hydrolysis of Dinitrogen Pentoxide Over the Temperature Range 214-263 K," Federal Aviation Administration Publication FAA-EE-81-3, Rockwell International Science Center, Los Angeles, CA (1981).
4. McArdle, J. V., and M. R. Hoffmann. "Kinetics and Mechanism of the Oxidation of Aquated Sulfur Dioxide by Hydrogen Peroxide at Low pH," *J. Phys. Chem.* 87:5425–5429 (1983).
5. Hoffmann, M. R., and J. G. Calvert. "Chemical Transformation Modules for Eulerian Acid Deposition Models, Vol. II, The Aqueous Phase Chemistry," National Center for Atmospheric Research, Boulder, CO (1985).
6. Martin, L. R., and M. W. Hill. "The Effect of Ionic Strength on the Manganese Catalyzed Oxidation of Sulfur (IV)," *Atmos. Environ.* 21:2267–2270 (1987).
7. Martin, L. R., and M. W. Hill. "The Iron-Catalyzed Oxidation of Sulfur: Reconciliation of the Literature Rates," *Atmos. Environ.* 21:1487–1490 (1987).

8. Huie, R. E., and P. Neta. "Rate Constants for Some Oxidations of S(IV) by Radicals in Aqueous Solutions," *Atmos. Environ.* 21:1743–1747 (1987).
9. Ross, A. B., and P. Neta. "Rate Constants for Reactions of Inorganic Radicals in Aqueous Solution," NSRDS-NBS 65, U.S. Department of Commerce, Washington, D.C. (1979).
10. Nair, S. K., and L. K. Peters. "Studies on Non-Precipitating Cumulus Cloud Acidification," *Atmos. Environ.* 23:1399–1423 (1989).
11. Chameides, W. L. "The Photochemistry of a Remote Marine Stratiform Cloud," *J. Geophys. Res.* 89:4739–4755 (1984).
12. Jacob, D. J. "Chemistry of OH in Remote Clouds and Its Role in the Production of Formic Acid and Peroxymonosulfate," *J. Geophys. Res.* 91:9807–9826 (1986).
13. Jacob, D. J., E. W. Gottlieb, and M. J. Prather. "Chemistry of a Polluted Cloudy Boundary Layer," *J. Geophys. Res.* 94:12,975–13,002 (1989).
14. Seigneur, C., A. M. Wegrecki, and L. W. Richards. "Mathematical Modeling of Cloud Chemistry in the California South Coast Air Basin," California Air Resources Board, Sacramento, CA (1989).
15. Seigneur, C., and A. M. Wegrecki. "Mathematical Modeling of Cloud Chemistry in the Los Angeles Basin," *Atmos. Environ.* 24A:989–1006 (1990).
16. Walcek, C. J., and G. R. Taylor. "A Theoretical Method for Computing Vertical Distribution of Acidity and Sulfate Production with Cumulus Clouds," *J. Atmos. Sci.* 43:339–355 (1986).
17. Tremblay, A., and H. Leighton. "A Three Dimensional Cloud Chemistry Model," *J. Climate Appl. Meteor.* 23:652–671 (1986).
18. Hales, J. M. "Comments on MAP3S/RAINE Precipitation Chemistry Statistics," *Atmos. Environ.* 17:1853–1854 (1983).
19. Pratt, G. C., and S. V. Krupa. "Seasonal Trends in Precipitation Chemistry," *Atmos. Environ.* 17:1845–1847 (1983).
20. Saxena, P., and C. Seigneur. "On the Oxidation of SO_2 to Sulfate in Atmospheric Aerosols," *Atmos. Environ.* 21:807–812 (1987).
21. Seigneur, C., and P. Saxena. "A Theoretical Investigation of Sulfate Formation in Clouds," *Atmos. Environ.* 22:101–115 (1988).
22. Seigneur, C., and P. Saxena. "A Study of Atmospheric Acid Formation in Different Environments," *Atmos. Environ.* 18:2109–2124 (1984).
23. Russell, A. G., G. J. McRae, and G. R. Cass. "The Dynamics of Nitric Acid Production and the Fate of Nitrogen Oxides," *Atmos. Environ.* 19:873–903 (1985).
24. Seigneur, C., P. Saxena, and V. A. Mirabella. "Diffusion and Reaction of Pollutants in Stratus Clouds: Application to Nocturnal Acid Formation in Plumes," *Environ. Sci. Technol.* 19:821–828 (1985).
25. Seigneur, C., P. Saxena, and P. M. Roth. "Computer Simulations of the Atmospheric Chemistry of Sulfate and Nitrate Formation," *Science* 225:1028–1030 (1984).
26. Saxena P., and C. Seigneur. "The Extent of Nonlinearity in the Atmospheric Chemistry of Sulfate Formation," *J. Air Pollut. Control Assoc.* 36:1151–1154 (1986).
27. Venkatram, A., and P. K. Karamchandani. "The ADOM II Scavenging Module: Incorporation of an Improved Cumulus Cloud Module," Electric Power Research Institute, Palo Alto, CA (1989).

28. Madronich, S. "Photodissociation in the Atmosphere. I. Actinic Flux and the Effect of Ground Reflection and Clouds," *J. Geophys. Res.* 92:9740–9752 (1987).
29. Seigneur, C., R. W. Bergstrom, C. D. Johnson, and L. W. Richards. "Measurements and Simulations of the Visual Effects of Particulate Plumes," *Atmos. Environ.* 18:2231–2244 (1984).
30. Schwartz, S. E. "MassTransport Considerations Pertinent to Aqueous Phase Reactions of Gases in Liquid Water Clouds," in *Chemistry of Multiphase Atmospheric Systems*. W. Jaeschlke, Ed. (Heidelberg: Springer-Verlag, 1986).
31. Seigneur, C., A. B. Hudischewskyj, J. H. Seinfeld, K. T. Whitby, E. R. Whitby, J. R. Brock, and H. M. Barnes. "Simulation of Aerosol Dynamics: a Comparative Review of Mathematical Models," *Aerosol Sci. Technol.* 5:205–222 (1986).
32. Saxena, P., A. B. Hudisckewskyj, C. Seigneur, and J. H. Seinfeld. "A Comparative Study of Equilibrium Approaches to the Chemical Characterization of Secondary Aerosols," *Atmos. Environ.* 20:1471–1483 (1986).
33. Pilinis, C., and J. H. Seinfeld. "Continued Development of a General Equilibrium Model for Inorganic Multicomponent Atmospheric Aerosols," *Atmos. Environ.* 21:2453–2466 (1987).

CHAPTER 10

Visibility in Urban Smog

Christine S. Sloane

Clear sunny days have aesthetic value for us all. They also have commercial value for tourism and real estate. Thus, the appearance of smoggy haze over cities is one of the most widely noted effects of air pollution. The task of protecting visual air quality requires an understanding of what visual haze is and where it comes from.

Over the past two decades, the need to understand visibility impairment has routinely been transformed into the seemingly straightforward task of evaluating the contribution of emission sources to the extinction of light in the atmosphere.[1-11] Our capacity to design effective control strategies, to mount effective monitoring efforts, and to predict with accuracy the visual effect of changing emission patterns hinges upon the relevance and accuracy of these light-extinction apportionments. This chapter points out situations where shortcomings of the apportionments can lead to ineffective regulatory strategies for urban visibility.

WHAT IS URBAN VISIBILITY?

The view of smog rolling into Riverside, CA, shown in Figure 1, illustrates the principal attributes of good and poor visibility: **visual range** (how far you can see), **sky coloration**, and **scene appearance** (the clarity of scenic details). Visual range is often defined as the greatest distance at which a black object can be seen silhouetted against the horizon sky. A newscaster who reports the visibility to be 10 or 15 miles is referring to the visual range determined by the National Weather Service at a nearby airport. There the weatherman goes outside and identifies the most distant building or terrain feature that he can see. In this case, visual range depends on more than air quality. It depends on the light source (sun angle, intervening clouds), on the reflectivity of the ground and visual targets, and on the eye-brain response of the observer.

To examine how pollutants affect visual range, standard viewing conditions are traditionally assumed. So-called standard viewing conditions are

Figure 1. Photograph of polluted air mass rolling into Riverside, CA.

illustrated in Figure 2. The sightpath is parallel to the ground through an evenly illuminated, uniform atmosphere, and a perfectly black target is viewed against a cloudless horizon sky. Visual range is traditionally defined as the distance at which the contrast between the radiance of the visual target and its background is only 2%. The contrast drops over distance because the radiance of light changes along the sightpath. When the distance from the target has increased to the point where the contrast is below 2%, the target is no longer considered to be visible.

The radiance of light seen by an observer consists of the radiance transmitted directly from the target plus path radiance, $I_p(R)$:

$$\underset{\text{Observed radiance}}{I(R)} = \underset{\text{Transmitted radiance}}{T(R) \cdot I(O)} + \underset{\text{Path radiance}}{I_p(R)}$$

R = target distance

Only the transmitted radiance carries information about the visual target.[12,13] The transmitted radiance decreases as light is scattered out of the sightpath or is absorbed. Path radiance consists of extraneous light scattered into the sightpath. As the sightpath lengthens, transmitted radiance decreases and path radiance increases, thereby decreasing visibility.

Transmittance is governed by the atmospheric extinction coefficient, B_{ext}, which is the sum of the scattering (s) and absorption (a) coefficients of particles (p) and gases (g) in the sightpath:

$$\text{Transmittance: } T(R) = \exp[-B_{ext} \cdot R] \quad (2)$$

$$\text{Extinction: } B_{ext} = B_{s,g} + B_{a,g} + B_{a,p} + B_{s,p}$$

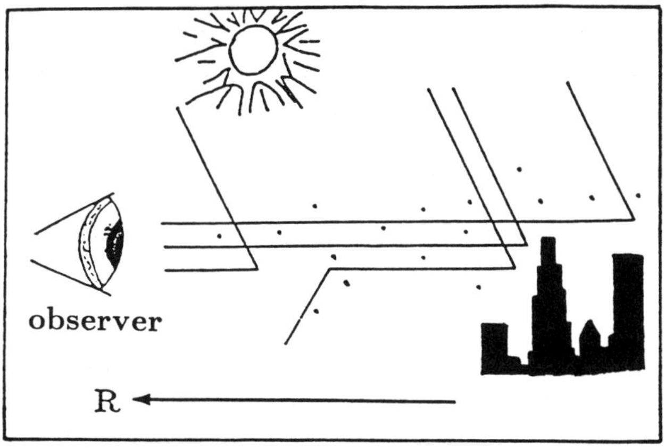

Figure 2. "Standard" viewing conditions: horizontal sightpath through a uniform atmosphere and over flat terrain. The visual target is perfectly black and is silhouetted against an optically thick horizon sky.

Clean air is responsible for gaseous light scattering, $B_{s,g}$; thus, clean air limits visibility. The main source of gaseous absorption $B_{a,g}$ is nitrogen dioxide, a regulated pollutant. In urban areas, light absorption by particles is primarily traced to soot. Airborne soil and dust contribute to absorption as well. Many pollutants contribute to particulate light scattering, $B_{s,p}$. Deciphering their contributions is the principal challenge in quantifying the sources of atmospheric light extinction. The extinction coefficient changes along a sightpath as pollutant concentrations vary. It is constant only in uniform, well-mixed air.

In urban areas, path radiance is usually the major cause of impaired visibility. It appears as the whitish haze superimposed on our field of view, as illustrated in Figure 3. Path radiance is increased by pollutants that scatter light into one's sightpath. Path radiance for a uniform sightpath can be predicted from the product of two terms:

$$\text{Path radiance: } I_p(R) = [1 - T(R)] \cdot I_e \qquad (3)$$

$$I_e = \text{equilibrium radiance (source function)}$$

The second term is the ultimate, or equilibrium, path radiance, which is reached as transmittance declines over a long uniform sightpath. This ultimate path radiance can be predicted from the atmospheric albedo, ω ($\omega = B_{scat}/B_{ext}$), and a complicated integral that depends on the viewing direction, Ω: $I_e(\Omega) = \omega \cdot \int P(\Omega,\Omega') I(\Omega')d\Omega'$.[14,15] $P(\Omega,\Omega')$ is the scattering phase function, the probability that extraneous light will be scattered into the sightpath. The albedo, ω, is simply the fraction of extinction that is due to scattering. Pollutants that absorb light lower the albedo, and also reduce $I(\Omega')$, the extraneous light incident on the sightpath. Thus, pollutants that absorb light lower I_e, the ultimate path radiance, and thereby limit path radiance under conditions of visibility impairment.

It is important to notice that extinction and path radiance are the two fundamental properties that control visibility. Both increase when light scattering is enhanced. However, they respond differently to light absorption.

VISUAL RANGE: A MISPERCEPTION

There are two principal reasons why visibility impairment has become associated with light extinction instead of path radiance. First, light extinction is easier to deal with. It is a property of the atmosphere alone, not of viewing conditions (sun angle, cloud cover, reflectivity of ground and viewed objects, distance along sightpath, etc.). Therefore, it is not necessary to distinguish pollutant effects from changes in viewing conditions. For the same reason, it is easier to measure light extinction. The measurement of extinction does not require a defined instrument orientation with respect to the sun, nor does it require a long unobstructed sightpath. The measurement of path radiance requires both.

Figure 3. Visual impairment from path radiance seen as a white veil superimposed on the field of view.

Viewing angle has a large influence on urban visibility. The change in path radiance with viewing angle is evident in the aerial photographs of urban wintertime haze shown in Figure 4. The first photograph was taken facing toward the sun. Light is strongly scattered in this forward direction, so the bright path radiance obscures the city below. The next photograph was taken 15 min later, after the plane flew around to the other side of the cloud. The sun is behind the photographer, and the path radiance from the haze, which is much less in the backward direction, is less intense (hence, darker) than that of the adjacent air. The same pollutants are present. A color photograph of this scene would show how "brown clouds" over some cities got their names. In this case, the difference between the appearance of the bright white haze and a brown cloud is caused by the change in viewing angle, not a change in pollutants.

The second reason for the focus on light extinction in visibility studies is historical. Before powerful computers were available to solve the radiative transfer equations that describe how light propagates in the atmosphere, mathematical simplifications that led to compact formulas were essential. Sometimes the lure of an easily understood formula obscured the limits of its relevance. For visibility, the classic simple formula is the oft-quoted Koschmieder equation:

$$\text{Visual range} = 3.9/B_{ext} \tag{4}$$

for the visual range of a perfectly black target (also, an optically thick sightpath to the target background and 2% contrast threshold for perception). The Koschmieder equation implies that visual range depends exclusively on the atmospheric extinction coefficient.

What's wrong with that? It conflicts with common experience. For example, sun glasses are worn to improve visibility even though they increase light extinction along the sightpath by absorption or reflection. Only when sun glasses are worn indoors (very low light conditions), or when they are scratched, causing annoying light to scatter into one's line of sight (path radiance), do they impair visibility.

The problem with the Koschmieder relationship is that it doesn't apply to common urban viewing conditions. A more general expression that does apply to urban viewing can be derived without the simplifications used to get the Koschmieder result. The observed contrast between a visual target (t) and the background (b) is

$$\text{Contrast:} \qquad C(R) = \frac{I_t(R) - I_b(R)}{I_b(R)} \tag{5}$$

R = target distance

This expression can be rewritten using Equation 1 for the target and background radiance:

$$C(R) = \frac{T(R) \cdot [I_t(0) - I_b(0)}{T(R) \cdot I_b(0) + I_p} = C(0) \cdot \frac{1}{1 + I_p/I_b(0) \cdot T(R)} \qquad (6)$$

which shows that if it weren't for path radiance (I_p), contrast would *not* decline with distance. Extinction reduces visual range when it is associated with an increase in path radiance (or light levels become so low that the scene appears black; i.e., $T \rightarrow 0$). The Koschmieder formula, which is equivalent to $C(R) = T \cdot C(0)$, applies only if $I_b(0)$, the radiance of the field of view adjacent to the visual target (the background), equals I_e. This condition does not describe common urban viewing conditions. Thus, the simple Koschmieder formula does not generally apply to realistic urban conditions, and visual range is *not* equally affected whether light extinction is caused by scattering or by absorption.

FINDING THE RESPONSIBLE POLLUTANTS: LIGHT EXTINCTION "BUDGETS"

The apportionment of the extinction to different pollutants is usually termed an extinction "budget". Since airborne particles dominate both the scattering and the absorption of light in most urban areas, the principal task is to relate the mass of the constituents of airborne particles to the atmospheric scattering and absorption coefficients. This is usually done by fitting optical measurements to measurements of the mass concentrations of the principal particle constituents, M_i. Regression techniques are used to estimate the fitting constants, a_i, in equations such as: $B_{ext} = \Sigma_i\, a_i \cdot M_i$.

In an extinction "budget", the contribution of each pollutant to B_{ext} is $a_i \cdot M_i$, the product of its regression coefficient and mass. Measurements generally include periods of excellent visibility and poor visibility. Therefore, extinction budgets usually describe average air quality rather than visibility impairment.

Typical extinction budgets[5,8] are shown in Figure 5 for western and eastern cities in the U.S. Western cities tend to experience wintertime "brown cloud" episodes, so investigative measurements are often made in the winter. The presence of agricultural ammonia emissions and low winter temperatures make ammonium nitrate a common constituent. Soot from diesel exhaust and wintertime wood-burning is also prominent. In western cities, light absorption often causes nearly a third of the atmospheric light extinction. Over half of the extinction is commonly caused by light scattering from carbonaceous material, ammonium sulfate (and the water associated with this hygroscopic salt), ammonium nitrate (and associated water), and by less prevalent constituents.

Conversely, summertime haze in eastern cities (east of the Mississippi River and one tier of states to the west) is generally dominated by acidic sulfates and water associated with acidic sulfates. Ammonium nitrate is not as prevalent in the east, because it is not as stable in acidic particles, nor in warm summertime temperatures.

Figure 4A. Aerial view of forward scattering from Denver haze.

Figure 4B. Aerial view of backward scattering from Denver haze.

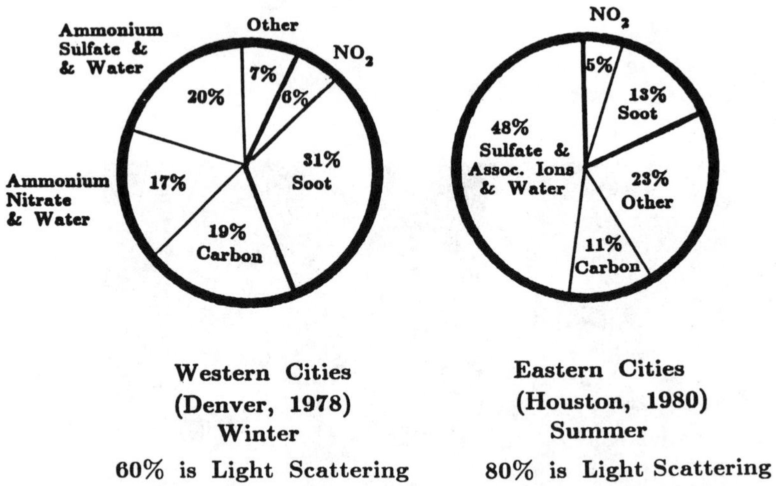

Figure 5. Typical extinction "budgets"

As discussed earlier, these extinction "budgets" can overestimate the role of light absorbers in the impairment of visual range. Scattering "budgets" are probably more relevant to path radiance. A comparison of extinction and scattering budgets, shown for Houston[8] in Figure 6, illustrates that the difference between the contribution of individual pollutants to extinction and to scattering is substantial. The difference between sources of extinction and scattering is greater for western cities where the concentration of soot is generally higher from wood burning. Control strategies focused on the reduction of extinction will be less efficient in reducing path radiance than strategies focused on light scattering.

Both **scattering and extinction "budgets"** can be misleading when used to predict how visual range might change in response to a change in

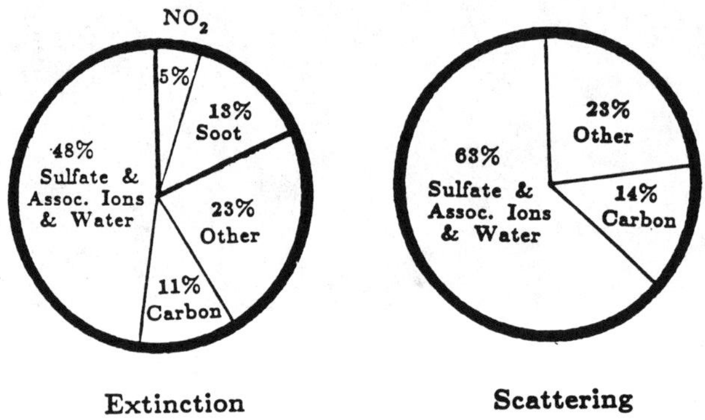

Figure 6. Comparison of scattering and extinction "budgets" for Houston, TX in 1980.[8]

pollutants.[16–18] For example, the scattering budget in Figure 6 suggests that light scattering could be reduced 48% by the removal of sulfates. This estimate is uncertain by at least 10%[17] for two reasons. First, light-scattering efficiencies (LSEs) are variables, but they are evaluated as the constants that fit measurements of mass, M_i, and extinction, B_{ext}: $B_{ext} = \Sigma_i a_i M_i$. An example of how LSEs vary is shown in Figure 7, which is a plot of LSE of pure soot, sulfate, and water particles shown as a function of particle size.[19] Hence, individual particulate constituents do not have constant visual impact per unit mass. Second, chemically pure particles are not found in the atmosphere, so the contribution of individual pollutants cannot be cleanly separated as it appears in a pie diagram of a budget.[16–18]

SKY COLORATION

Sky coloration is perhaps the most apparent attribute of visibility impairment after visual range. As with visual range, pollutants that scatter light have a different effect on sky coloration than pollutants that absorb light. For many years it was thought that sky discoloration was caused only by pollutants that absorb light, such as nitrogen dioxide (NO_2) gas, which absorbs blue light preferentially in the visible spectrum. That causes atmospheric light extinction (hence, light transmittance) to vary with wavelength differently than it would

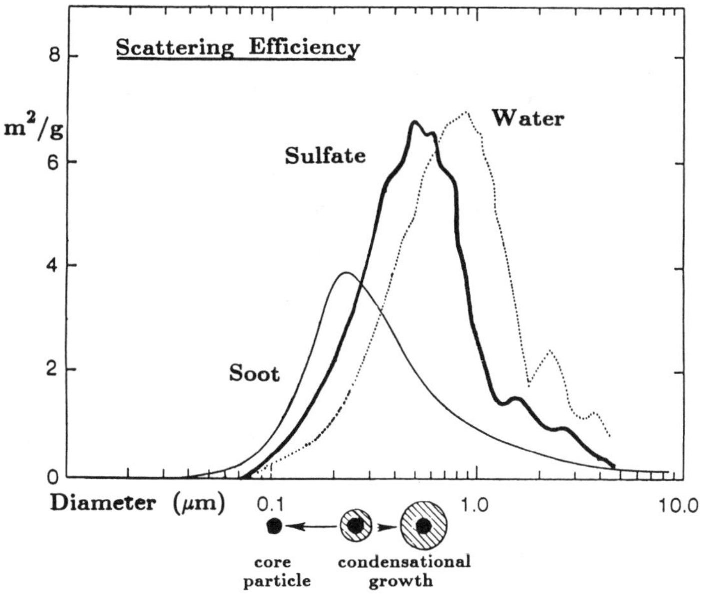

Figure 7. Variability of light-scattering efficiencies (light-scattering coefficient per unit mass) for prevalent constituents of urban fine particles. Formation of multicomponent particles by condensation is illustrated.

in clean air. However, research has shown[14,15,20,21] that light scattering by airborne particles causes sky coloration as well.

Sky coloration results from the perceived spectral distribution of path radiance along unobstructed sight paths that extend to outer space ($I(0) = 0$). The path radiance contributed by a uniform polluted layer within the atmosphere is described by Equation 3 where $I_e(\Omega) = [B_{scat}/B_{ext}] \cdot \int P(\Omega,\Omega')I(\Omega')d\Omega'$ for viewing direction Ω. Sky coloration varies with direction of view because the scattering phase function, $P(\Omega,\Omega')$, and extraneous light, $I(\Omega')$, are not the same for all wavelengths of light and all scattering angles.[14,21] Another factor that affects sky coloration is the wavelength dependence of the albedo $[B_{scat}/B_{ext}]$.[20]

Direct sunlight and light that has undergone multiple scattering within the atmosphere and with ground surfaces are the source of the extraneous light, $I(\Omega')$, that ultimately scatters into one's line of sight as path radiance. Multiple scattering within the atmosphere produces inseparable couplings between the visual impacts of individual pollutants. Hence, the sky coloration caused by individual emission sources can be evaluated only in the context of the other contributing sources.

As a result of these interwoven wavelength dependences, the path radiance produced by urban smog can have a different spectral distribution and intensity (hence, perceived color) from that of natural skylight produced by clean air. Differences between the sky coloration caused by airborne particles that scatter light and the coloration caused by particles and gases that absorb light have been explored using mathematical models that simulate how radiation propagates within a smog layer.[14,15,20,21] Calculations of the spectral distribution of light seen by an urban pedestrian within a smog layer generally show that the highest nitrogen dioxide levels currently reached in U.S. cities can produce a perceptible brownish-yellow coloration. However, comparable coloration can be produced by light scattered from airborne particles. Broadband absorbers, like soot, tend to diminish the color saturation.

In some instances, a brownish-yellow coloration appears from human chromatic adaptation. For example, an achromatic (i.e., grayish-white) field adjacent to a bright blue field can appear to have a brownish cast to an observer whose color response is saturated by the adjacent light field. This viewing condition occurs[22] when a smog layer containing strong achromatic absorbers, such as soot, is observed from where the field of view includes both the smog layer and brighter, bluer sky above the layer. Thus, even though a spectrometer measurement may show that the spectral distribution of the smog has an achromatic tone,[22] it may appear brownish-yellow.

SUMMARY

Visual range, sky discoloration, and scenic clarity are readily apparent attributes of urban visibility. Because visual range is the most easily understood and quantified attribute of visual impairment, it has become the focus of efforts to improve visibility. Strategies to protect and improve urban visibility depend on accurate and reliable accountings of the impact of different emission sources on visual range.

Considerable progress has been made in identifying the sources of urban visibility impairment since the inauguration of the Clean Air Act.[23] Computer models that predict atmospheric light intensities for different pollutant types and concentrations are available.[24] Shortcomings remain in the modeling of aerosol size distributions and the heterogeneity of like-sized particles.

The concern raised here is the tendency to assess visibility impact in urban areas in terms of atmospheric light-extinction. The appealing simplicity of the Koschmieder formula and the comparative experimental ease of light-extinction measurements are probably the two primary reasons for the focus on light extinction, rather than path radiance.

REFERENCES

1. Friedlander, S.K. *Smoke, Dust and Haze* (New York: John Wiley & Sons, Inc., 1977).
2. White, W.H., and Roberts, P.T., "On the Nature and Origins of Visibility-Reducing Aerosols in the Los Angeles Air Basin," *Atmos. Environ.* 11: 803–812 (1977).
3. Cass, G.R., "The Relationship Between Sulfate Air Quality and Visibility in Los Angeles," *Atmos. Environ.* 13: 1069–1084 (1979).
4. Ouimette, J.R., and Flagan, R.C., "The Extinction Coefficient of Multicomponent Aerosols," *Atmos. Environ.* 16: 2405–2419 (1982).
5. Groblicki, P.J., Wolff, G.T., and Countess, R.J., "Visibility-Reducing Species in the Denver Brown Cloud. I. Relationships Between Extinction and Chemical Composition," *Atmos. Environ.* 15: 2473–2484 (1981).
6. Shah, J.J., Watson, J.G., Cooper, J.A., and Huntzicker, J.J., "Aerosol Chemical Composition and Light Scattering in Portland, Oregon: the Role of Carbon," *Atmos. Environ.* 17: 386–389 (1983).
7. Wolff, G.T., Ferman, M.A., Kelly, N.A., Stroup, D.P., and Ruthkosky, M.S., "The Relationships Between the Chemical Composition of Fine Particles and Visibility in the Detroit Metropolitan Area," *APCA J.* 32: 1216–1220 (1982).
8. Dzubay, T.G., Stevens, R.K., and Lewis, C.W., "Visibility and aerosol composition in Houston, Texas," *Environ. Sci. Technol.* 16: 514–525 (1982).
9. Vossler, T.L., and Macias, E.S., "Contribution of Fine Particle Sulfates to Light Scattering in St. Louis Summer Aerosol," *Environ. Sci. Technol.* 20: 1235–1243 (1986).
10. Einfeld, W., Dattner, S., and Zimmerman, K., "The Dallas Winter Visibility Study," Sandia Report SAND88-1282, Sandia National Laboratories, Albuquerque, NM (1988).
11. Richards, L.W., Sloane, C.S., Watson, J.G., and Chow, J.C., "Extinction Apportionment for the Denver Brown Cloud," Trans. AWMA Visibility Specialty Conf., Air and Waste Management Association, Pittsburgh, PA (1990).
12. Malm, W. "Considerations in the Measurement of Visibility," *APCA J.* 29: 1042–1052 (1979).

13. Richards, L.W., "Sightpath Measurements for Visibility Monitoring and Research," *JAPCA* 38:784–791 (1988).
14. Husar, R.B., and White, W.H., "On the Color of the Los Angeles Smog," *Atmos. Environ.* 10: 199–204 (1976).
15. Richards, L.W., Bergstrom, R.W., and Ackermann, T.P. "The Optical Effects of Fine-Particle Carbon on Urban Atmospheres," *Atmos. Environ.* 20: 387–396 (1986).
16. Sloane, C.S., "Effect of Composition on Aerosol Light Scattering Efficiencies," *Atmos. Environ.* 20: 1025–1037 (1986).
17. Sloane, C.S., "Forecasting Visibility Impairment: a Test of Regression Estimates," *Atmos. Environ.* 22: 2033–2045 (1988).
18. White, W.H., "On the Theoretical and Empirical Basis for Apportioning Extinction by Aerosols: a Critical Review," *Atmos. Environ.* 20: 1659–1669 (1986).
19. Faxvog, F.R., and Roessler, D.M., "Carbon Aerosol Visibility vs. Particle Size Distribution," *Appl. Opt.* 17: 2612–2616 (1978).
20. Sloane, C.S., "The Contribution of NO_2 and Soot to the Discoloration of Urban Skies. I. Single Scattering," Trans. APCA Visibility Specialty Conf., 434–452, APCA, Pittsburgh, PA (1987).
21. Sloane, C.S., "The Contribution of NO_2 and Soot to the Discoloration of Urban Skies. II. Multiple Scattering," *Atmos. Environ.* 22: 2021–2031 (1988).
22. Waggoner, A.P., Weiss, R.E., and Ahlquist, N.C., "The Color of Denver Haze," *Atmos. Environ.* 17: 2081-2086 (1983).
23. Sloane, C.S., and White, W.H., "Visibility: an Evolving Issue," *Environ. Sci. Technol.* 20: 760–766 (1986).
24. Models currently available include NOAA Monte Carlo code, NASA doubling and adding code, GMR analytical multiple scattering code.

Index